AGROFORESTRY: REALITIES, POSSIBILITIES AND POTENTIALS

Agroforestry: Realities, Possibilities and Potentials

edited by

HENRY L. GHOLZ

Department of Forestry, University of Florida
Gainesville, Florida, USA

1987 **MARTINUS NIJHOFF PUBLISHERS**
a member of the KLUWER ACADEMIC PUBLISHERS GROUP
DORDRECHT / BOSTON / LANCASTER
in cooperation with ICRAF

Distributors

---

for the United States and Canada: Kluwer Academic Publishers, P.O. Box 358,
Accord Station, Hingham, MA 02018-0358, USA
for the UK and Ireland: Kluwer Academic Publishers, MTP Press Limited,
Falcon House, Queen Square, Lancaster LA1 1RN, UK
for all other countries: Kluwer Academic Publishers Group, Distribution Center,
P.O. Box 322, 3300 AH Dordrecht, The Netherlands

Library of Congress Cataloging in Publication Data

ISBN 90-247-3591-2 (paperback)
ISBN 90-247-3590-4 (hardbound)

TABLE OF CONTENTS

ACKNOWLEDGEMENTS

A number of individuals were instrumental in providing support for this production, physically, spiritually and financially.

Inspiration derived from discussions with Dr. Hugh Poponoe, Director of the Center for Tropical Agriculture, encouraged by Dr. Loukas Arvanitis, then the Acting Chairman of the Department of Forestry. Dr. Katherine Ewel established many of the lines of communication with eventual participants and many others who, for various reasons, could not participate.

Financial support was provided in many pieces by the Office of Academic Affairs, Division of Sponsored Research, Center for Tropical Agriculture, School of Forest Resources and Conservation, Department of Forestry, University Committee on Public Functions and Lectures, Center for Latin American Studies, Center for African Studies, Department of Soil Science, Women in Agricultural Development (WIAD) Program, and the Amazon Research and Training Program, all of the University of Florida. In some cases, partial support was provided indirectly from sources outside the University of Florida, without which this international gathering would not have been possible.

Dr. Katherine Ewel, Dr. Roger Webb, Dr. William McFee, Dr. Marianne Schmink, Dr. John Robinson, Dr. Alan Long, Mr. Pierre Berner, and Dr. Robert Schmidt provided reviews of at least one manuscript each, an invaluable contribution.

Editorial assistance in indexing was provided by Nancy Dohn Gholz, and in typing by Ms. Cathy Ritchie. Of course, I assume final responsibility for the editorial quality of this proceedings.

Finally, I greatly appreciate the opportunity to personally interact with the illustrious contributors to this series during their visits to Gainesville. Their support and encouragement for this endeavor was continuous. The new perspectives they imparted and the experiences they shared have altered permanently my views of land use and opened my eyes to a host of research possibilities and opportunities for international cooperation in the area of agroforestry.

1. INTRODUCTION

H. L. Gholz, Associate Professor, Department of Forestry, University of Florida, Gainesville, FL 32611.

Agroforestry is not a new concept. Trees, crops and animals have traditionally been raised together on small farms throughout the world. "Farm forestry" used to be a required course at many U.S. forestry schools and still is at a few. The demise of this concept in the temperate zone seems to have followed the demise of the small "family farm," as the trees, crops and animals have become separately managed at large scales in "modern" agriculture and forestry. Interest and expertise has waned as practitioners, scientists and educators focus on their own specific commmodity and research area.

The recent resurgence of the concept derives primarily from the recognized failure of large agriculture and forestry monocultures in the lesser developed world, primarily in the tropics. In addition, the rate of tropical forest destruction and loss of the multitude of natural products they yield, has raised an alarm worldwide. It is highly unlikely that mixed cultures will replace highly productive and profitable monocultures anywhere in the world as long as fossil fuel energy is abundant. However, in areas where fossil fuel subsidies are not possible, where soils are marginal or have been degraded, or where trees are now absent and their products sorely missed, increasing reliance on tree/crop/animal mixtures is inevitable.

In this context, a series of seminars was held at the University of Florida between October 1985 and June 1986 to foster a sustained, university-wide discussion of the general topic of "agroforestry." Although the university has separate programs in agriculture, forestry and the socio-economic aspects of natural resource use, historically there have been few attempts at integration. Furthermore, few faculty or administrators knew what "agroforestry" is, much less what research has been done, or had thought about how social and biological scientists could or should cooperate. Is agroforestry a useful and appropriate scientific area for university scientists to become involved in?

Some questions were answered relatively easily. Agroforestry has been defined by the International Council for Research in Agroforestry (ICRAF) as: "...a collective word for all land-use practices and systems in which woody perennials are deliberately grown on the same land management unit as annual crops and/or animals." Even though some may criticize this definition as being too broad and all-inclusive, I have, in retrospect, found it to be a very useful one. The inclusion of "deliberately" eliminates many naturally or casually occurring mixes of trees, crops and/or animals, and the definition allows for both traditional and modern variations.

Classifying the myriad observable or hypothetical agroforestry systems is more problematic, and some of the participants in this series disagreed quite strongly as to how a suitable scheme should appear. Some schemes have already been produced (e.g., 1, 5), and Drs. Nair, Tejwani, Rocheleau, Lewis and Budowski each contributed greatly to my awareness of the issues through their original submissions. Some of the best schematic

Table 1. Some examples of prominent agroforestry systems and practices in the developing countries based on the results of ICRAF's Agroforestry Systems Inventory Project (4).

| TYPE OF SYSTEM | | EXAMPLE |
Major System	Sub-System/Practices	South Pacific
AGROSILVICULTURAL SYSTEMS	Improved Fallow (in shifting cultivation areas)	-
	The Taungya System	(e.g., Taro with Cedrella and Anthocephalus trees)
	Tree Gardens	Involving fruit trees
	Hedgerow Intercropping (Alley Cropping)	
	Multipurpose Trees and Shrubs on Farmlands	Mainly fruit or nut trees (e.g., Canarium, Pometia, Barringtonia, Pandanus, Artocarpus altilis)
	Crop Combination with Plantation Crops	Plantation crops and other multipurpose trees (e.g., Casuarina and coffee in the highlands of PNG; also Gliricidia and Leucaena with cacao)
	Agroforestry Fuelwood Production	Multipurpose fuelwood trees around settlements
	Shelterbelts, Windbreaks, Soil Conservation Hedges	Casuarina oligodon in the highlands as shelterbelts and soil improvers
SILVOPASTORAL SYSTEMS	Protein Bank (Cut-and-carry Fodder Production)	Rare
	Living Fence of Fodder Trees and Hedges	Occasional
	Trees and Shrubs on Pastures	Cattle under coconuts, pines and Eucalyptus deglupta
AGROSILVOPASTORAL SYSTEMS	Woody Hedges for Browse, Mulch, Green Manure, Soil Conservation, etc.	Various forms; Casuarina oligodon widely used to provide mulch and compost
	Home Gardens (involving a large number of herbaceous and woody plants with or without animals)	Several types of homegardens and kitchen gardens
OTHER SYSTEMS	Agro-Silvo-Fishery ('Aquaforestry')	
	Various forms of Shifting Cultivation	Common
	Apiculture with Trees	Common

[1] For further information contact: Dr. P. K. R. Nair, Coordinator, AFSI, ICRAF, P. O. Box 30677, Nairobi, Kenya.

Table 1. (Continued)

	EXAMPLES		
	Southeast Asia	South Asia	Middle East and Mediterranean
	Forest villages of Thailand; various fruit trees & plantation crops used as "fallow" species in Indonesia	Improvements to shifting cultivation; several approaches (e.g. in the north-eastern parts of India)	
	Widely practised; forest villages of Thailand are improved forms	Several forms, several names	
	Dominated by fruit trees	In all ecological regions	The Dehesa system; "Parc arboree"
	Extensive use of Sesbania grandiflora, Leucaena leucocephala and Calliandra callothyrsus	Several experimental approaches (e.g., conservation farming in Sri Lanka)	
	Dominated by fruit trees; also Acacia mearnsii-cropping system, Indonesia	Several forms both in low-lands and highlands (e.g., hill farming in Nepal; 'Khejri' - based system in the dry parts of India)	The oasis system; crop combinations with the Carob tree; the Dehesa system; irrigated systems; olive trees + cereals
	Plantation crops + fruit trees; smallholder systems of crop combinations with plantation crops; plantation crops with spice trees	Integrated production systems in smallholdings; shade trees in plantations; other crop mixtures including various tree spices	Irrigated systems; olive trees + cereals
	Several examples in different ecological regions	Various forms including some forms of social forestry	
	Terrace stabilization in steep slopes	Use of Casuarina spp. as shelterbelts; several windbreaks	Tree spices for erosion control
	Very common, especially in highlands	Multipurpose fodder trees on or around farmlands especially in highlands	
	Leucaena, Calliandra, etc., used extensively	Sesbania, Euphorbia, Syzigium, etc., common	
	Grazing under coconuts and other plantations	Several tree species used very widely	Very common in the dry regions; the Dehesa system
	Various forms	Various forms especially in lowlands	
	Very common; Java home gardens often quoted as examples; involving several fruit trees	Common in all ecological regions; usually involving fruit trees	The Oasis system
	Silviculture in mangrove areas; trees on bunds of fish-breeding ponds	Occasional	
	e.g., swidden farming	Very common; various names	
	Common	Common	Common

Table 1. (Continued)

| | EXAMPLES | |
East and Central Africa	West Africa	American Tropics
Improvements to shifting cultivation (e.g., gum gardens of the Sudan)	Acioa barterii, Anthonontha macrophylla, Gliricidia sepium, etc., tried as fallow species	Several forms
The 'Shamba' system	Several forms	Several forms
		e.g., 'Paraiso Woodlot' of Paraguay
The corridor system of Zaire	Experimental systems on alley cropping with Leucaena and other woody perennial species	Experimental
Various forms; the Chagga system of Tanzania highlands; the Nyabisindu system of Rwanda	Acacia albida-based food production systems in dry areas; Butyrospermum + Parkia systems; "Parc arboree"	Various forms in all ecological regions
Integrated production; shade trees in commercial plantations; mixed systems in the highlands	Plantation crop mixtures; smallholder production systems	Plantation crop mixtures; shade trees in commercial plantations; mixed systems in smallholdings; spice trees; babassu palm-based systems
Various forms	Common in the dry regions	Several forms in the dry regions
The Nyabisindu system of Rwanda	Various forms	Live fences, windbreaks especially in highlands
Very common	Very common	Very common
Very common in all ecological regions		Very common in the highlands
The Acacia-dominated system in the arid parts of Kenya, Somalia and Ethiopia	Cattle under oil palm; cattle and sheep under coconut	Common in humid as well as dry regions (e.g., grazing under plantation crops in Brazil)
Common; variants of the "Shamba" system	Very common	Especially in hilly regions
Various forms (e.g., the Chagga homegardens; the Nyabisindu system)	Compound farms of humid lowlands	Very common in the thickly populated areas
Very common	Very common in the lowlands	Very common in all ecological regions
Common	Common	

illustrations I have seen were submitted by P. K. R. Nair as a result of his synthesis work at ICRAF, and with his permission, I present one here as Table 1. This scheme provides examples of agroforestry systems in use in various geographic regions of the world, and includes many synonyms for the same system as used in different areas. Essentially, all the chapters in this volume, then, address one or more of these systems and Table 1 provides a very useful method for cross reference and comparison.

There are still important areas in the classification of agroforestry systems that need interpretation. For example, should slash and burn (or swidden) agriculture be included in agroforestry? My interpretation would indicate "perhaps". In his introductory chapter, Vergara includes such "rotational" systems in agroforestry, and certainly if the objective of the swiddeners is to eventually return to the same area after a sufficient fallow period, I would agree. Or, as Padoch and de Jong document, if the "fallow" is actually managed to some extent or "improved", then this becomes a deliberate agroforestry system. But, if the swiddener's purpose is to clear forest to establish land tenure for speculation, to convert the area to permanent pasture, or to abandon the area and move on in true frontier fashion, then this activity, in my mind, is destructive and should not be regarded as agroforestry.

Another area of much debate concerns "Taungya", eloquently discussed in a Nigerian history context by Lowe, and mentioned frequently by others. Is Taungya only a purposeful exploitation of peasant workers to establish tree plantations? Lowe concludes that its use must be kept in perspective. If a government needs and wants to regenerate forest areas, this may be a mutually beneficial approach: farmers continue to get newly cleared land and the government gets its trees planted. But, even in the most positive context, Lowe concludes that it "depends on the existence of a peasant population in a substantially subsistence economy at a low standard of living."

It seems to be this unavoidable and integral human presence that differentiates agroforestry from modern agriculture or forestry. Can agroforestry, then, include larger, commercial operations? Certainly the cattle/pine systems described by Lewis and Pearson should qualify, although the human component is much less intimate than in the home gardens described by Tejwani. Altieri and coauthors provide data on insect interactions that have implications for pest management in all agroecosystems, large or small.

The seminar series was definitely successful in fostering a sustained discussion of agroforestry, and, in fact, has formed the basis for the development of an agroforestry program at the University of Florida. Over 150 individuals from 13 campus units attended at least one seminar and 29 students from 9 units registered for credit. The average attendance was 40 and the range from 25 to over 100. The diversity and depth of interest was startling. However, although interest is now high among faculty and students, it is diffuse. Few individuals, except the registered students, attended a majority of the seminars, and no clear concensus for action emerged. Effective multi-disciplinary research, as called for by the definition of agroforestry, makes many academicians squeamish, and certainly does not occur spontaneously.

Obviously, there is no reason that most all existing plant, soil and systems research couldn't be focused on agroforestry plants, soils or systems--I have failed to identify anything particularly unique about these resources in this context. In fact, Huxley (3) and Nair (6) in previous volumes have explored plant and soil research as it might apply to the analysis of agroforestry systems, and ecologists are used to examining multi-species arrays, natural or otherwise (e.g., 2).

What important areas of research, then, can be identified? The hope is that these papers will promote a discussion of the possibilities, which are virtually endless as a result of the newness of the field to science. The papers here essentially pose a tremendous set of hypotheses concerning agroforestry. Tejwani questions whether there has been enough research done in India for science to have contributed at all yet to the improvement of traditional or development of new agroforestry systems. Anderson proposes a novel pattern of nutrient cycling while examining the use of two native palm-dominated forests. Both Budowski and Brewbaker challenge us with new views and uses for "trees" and how they may fit into small, or large, farming systems. And, with a very different perspective, Eisenberg and Harris remind us how extremely important wild animals are to small farmers, who are often hunters as well. They also pose the intriguing question of the potential role of agroforestry as a land use of intermediate intensity (between natural forest management and plantation agriculture and forestry), in designing landscapes that maximize biodiversity while providing more immediately useful products to humans as well. Finally, Rocheleau derives and illustrates a program for the incorporation of the farmers' perspectives into research and development programs in agroforestry, highlighting the important role of the social sciences in evaluating agricultural land use alternatives on small farms.

There is no question that most "research" in agroforestry so far has been highly descriptive and/or empirical. With a wide-open expanse of possibilities, a greater focus on processes and principles is essential. Description and empiricism must also continue, but, if conducted by themselves, only slow progress will be made. It seems to me that university researchers could contribute most in this regard. Hopefully, the papers presented here, and the extensive literature assembled, will encourage discussion of the important subject of agroforestry, its realities, potentials and the research opportunities it presents.

REFERENCES

1.   Combe, J. 1982. Agroforestry techniques in tropical countries: potential and limitations. Agrofor Syst 1:13-28.
2.   Ewel, J. J. 1986. Designing agricultural ecosystems for the humid tropics. Ann Rev Ecol Syst 17:245-71.
3.   Huxley, P. A. (ed). 1983. Plant research and agroforestry. Intern Council Res Agrofor (ICRAF), Nairobi, Kenya. 617 pp.
4.   Nair, P. K. R. 1985. Classification of agroforestry systems. Agrofor Syst 3:97-128.
5.   Nair, P. K. R. (ed). 1984. Soil productivity aspects of agroforestry. Intern Council Res Agrofor (ICRAF), Nairobi, Kenya.

2. AGROFORESTRY: A SUSTAINABLE LAND USE FOR FRAGILE ECOSYSTEMS IN THE HUMID TROPICS

Napoleon T. Vergara

Research Associate, Environment and Policy Institute, East-West Center, Honolulu, Hawaii, U.S.A.

2.1. ABSTRACT

Many new farms carved out of upland forests and cropped monoculturally and intensively with annual plants are difficult to maintain in a productive state because of the climatic, topographic, and socioeconomic conditions which lead to abusive use and eventual site degradation. Sustaining these new farms with expensive chemical inputs seems infeasible because of the meager capital resources of upland farmers. The most viable alternative for maintaining productivity is the application of low-cost agroforestry land-use systems which promote soil conservation and minimize nutrient losses.

Traditional swidden agroforestry, which draws its sustainability from the rehabilitative effects of forest fallow, can remain sustainable only where population pressure is sufficiently light to allow long "rest" or fallow periods, or where efficient fallow species can rejuvenate the site over shorter periods. Where population densities preclude fallowing, however, the better system appears to be integral agroforestry, which involves simultaneous and continuous cropping with annuals and perennials. Integrating agroforestry through spatial manipulation of the crop components and selection of perennial species which maximize productivity and site protection promotes sustainability.

2.2. INTRODUCTION

2.2.1 Hunger Amid Productivity: A Paradox in the Tropics

As an ecological zone, the humid tropics has an immense potential for high biological productivity. Precipitation, ranging from 1,800 mm to 4,600 mm per year, exceeds evapotranspiration and nets sufficient moisture in the soil to nurture plants. A substantial amount of solar radiation is received to boost photosynthesis since diurnal hours are almost uniformly long and insolation occurs most days of the year. There are no extreme seasonal temperature variations to cause periodic plant dormancy and considerable nutrient inputs are provided through rainfall and litterfall. The vast species diversity and the luxuriant growth of tropical rainforests are indicative of this productive potential.

Paradoxically, however, most of the countries that are socioeconomically depressed and whose populations at times suffer from food shortages and malnutrition are located in the potentially productive humid tropics. Agricultural productivity has apparently been declining through inappropriate or abusive land use, and aggregate outputs seem unable to keep up with rapidly escalating population (22).

2.2.2 Fragility of Tropical Ecosystems

Humid tropical ecosystems are known to be fragile, i.e., prone to environmental degradation and loss of productive capacity when disturbed by human activities. Ironically, the inherent susceptibility of tropical upland areas[1] to ecological deterioration is brought about by the very same factors that favor rapid plant growth and productivity. High temperatures and humidity, which hasten the breakdown of parent material and decomposition of organic matter, also increase the likelihood of nutrient loss through leaching and runoff. Heavy precipitation can very easily turn into torrential rains that erode nutrient-bearing soil and leach away soluble nutrients, thereby quickly impoverishing the site. Intense solar radiation can also result in excessive moisture and temperature stress among plants and may render some cleared sites unsuitable to planting.

2.3. STRATEGIES FOR MAINTAINING PRODUCTIVITY

With temperature and moisture conditions generally favorable, nutrients become the most important limiting factor in agro-based production schemes in the humid tropics. The ability of land managers to maintain soil nutrients at reasonably high levels at little or no cost is the key to success in sustaining productivity.

2.3.1 Nutrient Flows in an Ecosystem

The nutrient budget of an ecosystem is never static. Nutrient inflow, as well as outflow, occurs continuously (Figure 1). In a long-undisturbed forest in the humid tropics, the quantity of inputs (A, C, D) equals approximately the magnitude of the losses (E, G, H) (5).

Natural nutrient inputs exist in three major forms: nutrient-bearing dust particles in the atmosphere that combine with rainfall; atmospheric nitrogen fixed in root nodules and brought to the surface soil as litter fall by plant species capable of biological nitrogen fixation; and leached nutrients entering as runoff from adjacent and higher-elevation areas.

Some transfers within the ecosystem may render some nutrients more or less available to certain plants. For instance, newly released nutrients from parent materials in the subsoil cannot be utilized by shallow-rooted plants that may have access to them only after they are absorbed by deep-rooted trees and translocated to the soil surface as litterfall or leached from leaves. Some nutrients taken up by perennials are stored in aboveground biomass and are recycled only when the tree dies and decomposes, or when it is cut and burned on site, as in slash and burn cultivation.

Nutrient losses from an undisturbed system can occur in the forms of erosion, leaching, and runoff, and through nutrient exports (i.e., leaves, flowers and fruits gathered by humans or eaten by animals on site but recycled as waste elsewhere). Apparently, in an undisturbed site, the natural nutrient inputs (and transfers from the subsoil) are greater than natural losses. The strongest empirical evidence for this is the fact that a sufficiently long forest fallow can bring the nutrient contents of a depleted swidden soil back to the original level (6, 2, 20, 28).

---

[1]"Uplands" are defined (19) as lands over 500 meters in elevation and above 20 percent in slope.

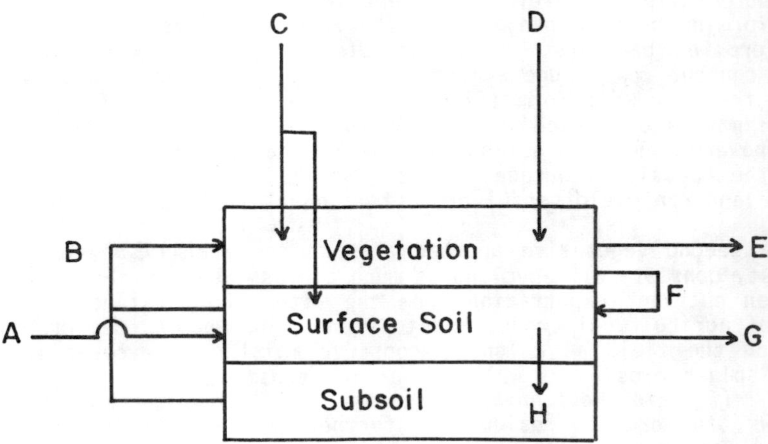

Figure 1.   Nutrient flows in an ecosystem.   A = nutrient inflows from adjacent sites, B = nutrient uptake by plants, C = nutrient input through rainfall, D = nitrogen input through biological fixation, E = nutrient drain through exportation, F = litterfall, G = nutrient loss through erosion and leaching, H = percolation.

When a natural forest ecosystem is transformed into an agroecosystem, the rate of nutrient drain increases considerably and usually exceeds the natural inputs by a substantial amount. The nutrient removals are in the forms of: (1) forest cutting and biomass exportation, as in logging prior to cropping; (2) accelerated erosion and runoff caused by exposure to the erosive forces of rain and wind, and loosening of soil surface through cultivation; and (3) excessive nutrient exportation through frequent harvest of annual crops. Unless some measures are quickly taken to narrow the widening gap between the huge nutrient losses and minimal natural inputs, the agroecosystem can become severely depleted.

2.3.2. <u>Some Approaches to Minimizing Nutrient Drains</u>

The possible ways of reducing erosion and nutrient losses from fragile upland sites may be grouped under mechanical or vegetative means. Terracing, contour ditching, and contour banking to slow runoff and reduce erosion are excellent examples of mechanical or engineering techniques. They are clearly effective as evidenced, for example, by the continued operation of centuries-old rice terraces in southern China, Bali (Indonesia), Nepal, and Ifugao (Philippines). At this time, however, they do not seem to be attractive alternatives to most farmers, as shown by the substantial expansion in upland populations unaccompanied by corresponding enlargement of terraced or contour-banked land. This lukewarm attitude towards the engineering approach is likely to have arisen because of the difficulty of terracing increasingly rough terrain among new upland farms, by the unsettled tenure over most upland areas, and by labor and capital deficiencies among marginal farmers which preclude the practice of labor- and capital-intensive conservation measures.

Vegetative approaches to soil conservation are of two main types. The first entails the careful placement of deep-rooted trees along contour lines, forming contour hedges to anchor and stabilize slopes and act as living erosion barriers. Soil particles eroded from the clear strips between contour rows lodge against the "live wall" such that, over time, natural terraces will form (Figure 2). One important advantage of this over the man-made terrace is that it involves comparatively less capital and manpower inputs to establish, operate, and maintain. In addition, unlike the "passive" terrace, the contour hedgerow continuously protects the site and can yield useful products such as fuelwood, fodder and green manure.

The second vegetative approach, which can readily be combined with the first, consists of recycling as much biomass as possible after harvest to lessen nutrient exportation from the site. For instance, leaves and stalks of agricultural crops, and tops and branches of tree crops can be spread on the field as mulch to conserve moisture, as cover to protect against splash erosion as well as surface erosion, and as recycled green manure. If laid horizontally against the upslope side of contour hedgerows, the organic residue can further increase the ability of such hedgerows to reduce erosion and runoff.

2.3.3. Some Means for Maximizing Nutrient Inflows

Nutrient inflows can be managed and enhanced to a certain extent. Little can be done to manipulate the quantity of precipitation and the amount of nutrient-bearing dust particles in the atmosphere that contribute to site nutrient budgets. One effective way of enhancing natural inputs, however, is through the introduction of N-fixing trees, such as those belonging to the legume, _Casuarina_, and _Alnus_ groups. With

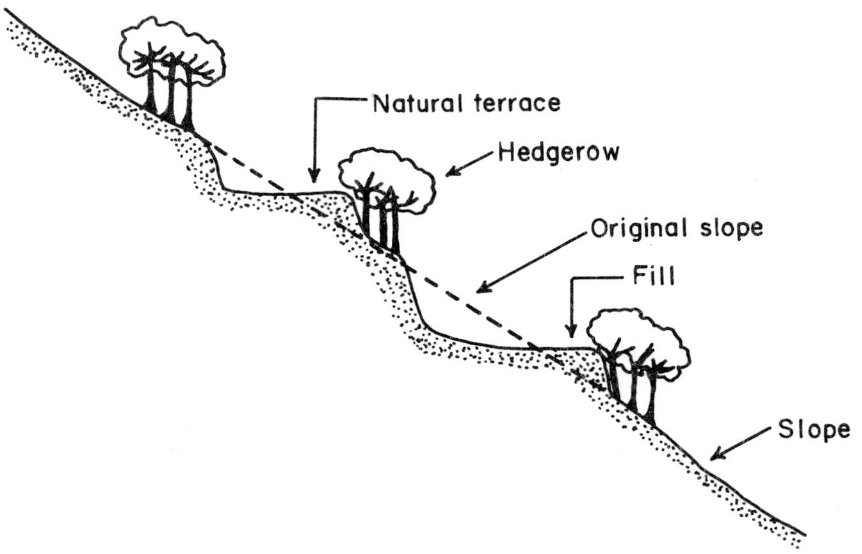

Figure 2. Natural terraces formed by contour hedges (24).

the assistance of microbial agents, such as <u>Rhizobia</u> and <u>Frankia</u>, these trees can supply nitrogen at almost no cost and on a continuing basis. Since this approach calls for human intervention in the selection of species with N-fixing capabilities, and in the inoculation of these species with N-fixing microbes, it falls under the category of "man-enhanced" natural inputs.

Applying chemical or inorganic fertilizers on depleted lands is one of the quickest ways or restoring or maintaining productivity. It is difficult to use this approach in the case of marginal upland farms, however. First, there is a need to replace not only the nutrients but also the soil lost through erosion. Inorganic fertilizers can add nutrients but cannot contribute to soil formation, unlike fallow vegetation, for example, which can accumulate organic matter from litter by as much as 5 to 17 tons per hectare per year (7, 21). Second, the high cost of chemical fertilizers, relative to the farmers' meager income, is prohibitive. Since most upland farmers have little or no cash incomes, this type of input is out of the question for most. For these reasons, the greatest hopes for regaining or maintaining productive capacity rests upon low-cost biological approaches such as the use of forest fallows or introduced N-fixing trees. Agroforestry, as a production system, can combine nutrient- and soil-conserving trees and soil-forming as well as soil-improving perennials with food crops, and therefore possesses the potential for achieving the desired sustainability in managed areas of the humid tropics (12, this volume).

2.4. AGROFORESTRY SYSTEMS: THEIR POTENTIAL FOR SUSTAINED YIELDS

As a land-use system, agroforestry is fairly known. It is sufficient to state that agroforestry is a system combining agricultural and tree crops of varying longevity (ranging from annual through biannual and perennial plants), arranged either temporally (crop rotation) or spatially (intercropping), to maximize and sustain aggregate yields. At times livestock and fish may be added as components (25).

2.4.1. <u>Swidden Cultivation: A Traditional Rotational Agroforestry System</u>

Swidden cultivation, also known as slash-and-burn or shifting cultivation, among other localized names, is referred to as rotational agroforestry because the perennial components (the forest fallow ranging from 8 to 40 years in duration) are rotated with the cultivation of annual food crops (cropped over a period of one to three years). A complete swidden cycle is the sum of the fallow and the cropping phases (Fig. 3).

Since the creation of national forestry authorities, swidden has been considered wasteful of timber resources and a key cause of deforestation and ecological degradation. These widely held views are based on readily visible evidence, including cutting and burning of large-size, commercially valuable timber; severe sheet and rill erosion during the cropping period; abandonment of heavily degraded sites to grass or bush vegetation; siltation of streams and lakes; and adverse impact on irrigation and lowland production. Where frequent burning of grass occurs, woody vegetation may be unable to emerge as fallow and the site remains an "unproductive" grassland. These negative views have motivated national and international agencies to launch vigorous campaigns to eradicate this form of land use (8).

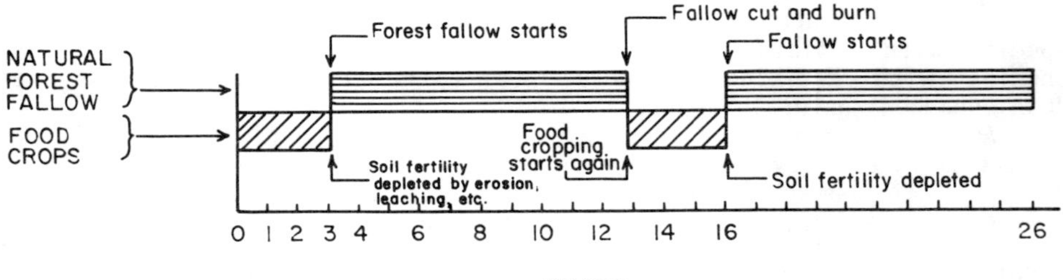

Figure 3. Swidden or rotational agroforestry cycle.

Ironically, it took social scientists like Conklin (6), rather than natural scientists to recognize the long-term sustainability of swiddens. Land-use specialists (foresters and agriculturists), usually confining their observations solely to the slashing and cropping phases, noted only the apparent destructive aspects of slash-and-burn, and they became adamantly opposed to this system. Long-term investigations, however, led to the "discovery" of the rejuvenative functions of the forest fallow and the restoration of soil nutrients and productivity to original levels (Fig. 4). This observation is strongly backed by empirical evidence that under sufficiently long fallows, swidden sites can become productive again and have been used by traditional farmers in repetitive cycles. The system has been practiced in the tropics for millenia (11) and is the outcome of centuries of trial-and-error practices by traditional farmers.

A traditional rotational agroforestry system usually starts with slashing and burning for quick transfer of nutrients from the forest biomass to the soil followed by the planting of food crops. Burning the slashed forest vegetation produces ashes and serves the added purpose of eliminating noxious weeds and harmful soil fauna, such as nematodes (28). The species established during the cropping phase differ from that of the original forest. Yet there is often much similarity in the wide diversity of species and plant sizes and in an often multilayered canopy. In the Western Province of the Solomon Islands, for instance, swidden farmers plant food crops ranging from creeping sweet potato (Ipomoea), low-level taro (Colocasia) to high-level cassava (Manihot); non-woody semi-perennials ranging from low-level sugarcane (Saccharum), mid-level papaya (Carica) to high-level banana (Musa); and woody perennials such as breadfruit (Artocarpus) and Sago palm (Metroxylon). This polycultural nature of swidden crops is further heightened by the availability of numerous varieties within each species. Solomons and New Guinea farmers, for example, have identified four varieties of sweet potato, three of yam, five of bananas, and two of sugarcane (United Nations University/East-West Center on-going research in the South Pacific , 1986). This diversity reflects deliberate efforts to: (1) insure against risks of crop failure and pest infestation, (2) to fill the various ecological niches for greater productivity per hectare per year, and (3) to widen the range of nutritional supply. A good case of risk minimization is demonstrated in Pacific swiddens where certain insects perforate the leaves of sweet potato so thoroughly that photosynthetic action is adversely affected and yields of that species are low but regular; none of the other intercropped

12

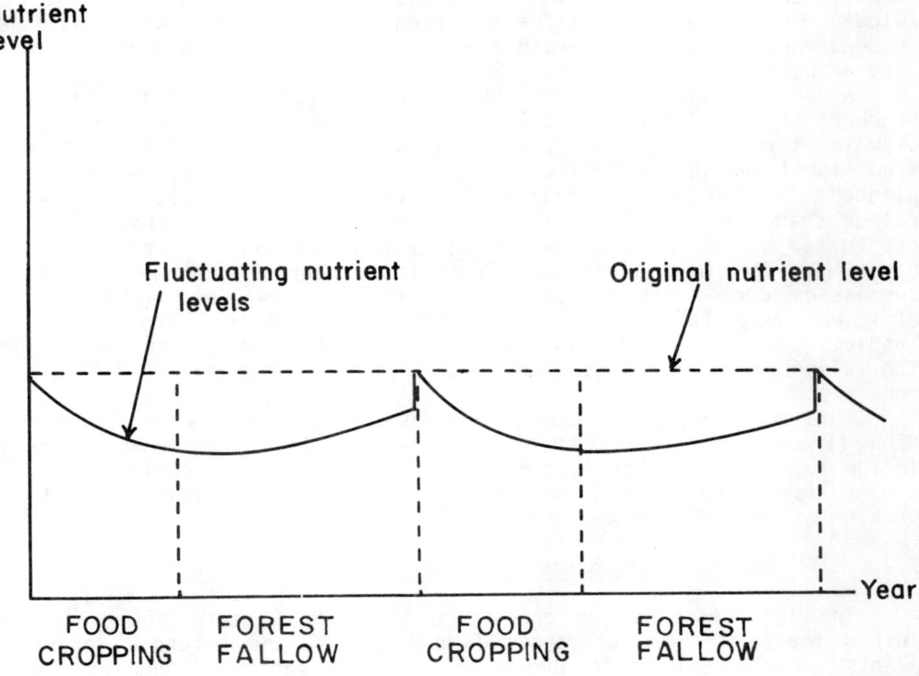

Figure 4. Hypothetical nutrient fluctuations during swidden cycle.

food plants are similarly affected so that the farmers' food flow is not threatened.

Since most swiddens are on slopes, rapid nutrient depletion can also occur through leaching and erosion. Combined with the intrusion of noxious weeds, these factors force the farmer to seek a new site after only one to three years of continuous cropping. Fallow sets in afterwards to start the process of site rehabilitation.

An abandoned swidden does not necessarily cease to yield products for the farmer as longer-lived food crops continue to produce (see Padoch, this volume). In Asia and the Pacific, farmers continue to harvest pineapple, banana, sugarcane, and papaya for about two to three more years after the site is left to fallow (United Nations University/East-West Center on-going research in the South Pacific, 1986). This long overlap between food production and fallow could abate the concern of many policy makers about the long withdrawal of fallow land from production, which is hardly affordable under currently heavy people pressure on shrinking land resources.

The relentless pressure on land is the main cause for the loss of sustainability of upland swidden cultivation. The opportunity to rotate fields diminishes, and farmers resort to lengthening cropping periods and shortening fallows until swiddens become almost settled cultivation. Longer cropping of hilly sites means more severe soil and nutrient depletion, which requires a longer fallow, but the abbreviated fallow is insufficient to restore productive capacity to the original level. A swidden farm under this situation progressively degenerates with each succeeding cycle until it reaches an irreversible level and is completely lost as wasteland.

A certain management technique has been developed to maintain swidden productivity even under short fallows. In Cebu, central Philippines, for example, farmers have modified the swidden system which differs from the traditional one in the following ways: (1) The forest fallow consists of planted nitrogen-fixing, fast-growing, coppicing tree species (_Leucaena_) rather than a mixture of "volunteer" species; (2) The fallow biomass is not burned but is cut and neatly laid along contours to serve as erosion bars; (3) The long and less efficient herb/grass/bush/tree vegetative succession during the natural fallow phase is replaced by a fast and efficient tree fallow system that immediately emerges from sprouts or coppice; and (4) Sprouts that emerge during the cropping phase or before the fallow period are periodically cut and applied as nitrogen-rich green manure and mulch.

This modification of traditional rotational forestry has enabled the Cebu farmers to lengthen the crop period to four years and reduce the fallow to six years. It is a substantial improvement over the usual 3:10 crop-to-fallow ratio in traditional systems and is a good answer to the problem of people pressure on shrinking land areas.

2.4.2. The Intercropping System: A "Modern" Type Agroforestry

The intercropping (as opposed to rotational) form of agroforestry infers the continuous presence of both annual and perennial groups of plants on the same site under any of the various types of spatial arrangements (Figure 5). One major advantage is its ability, if properly implemented, to continuously yield food crops and tree products from the same site without an intervening "idle" fallow period. An important reason for this unbroken productivity is the continuous rather than intermittent roles of the interplanted trees in reducing nutrient drain and maximizing nutrient inputs. Because this concept varies dramatically from the traditional agroforestry or swidden system, it is often called "modern" agroforestry.

Natural forests, with trees randomly placed in the landscape, reduce erosion by (1) anchoring the weaker surface soil horizon on bedrock crevices and by binding together the surface soil with their woody and interlocking root systems (14, 15, 27); (2) providing protection against erosion with a "triple armor" consisting of the forest canopy, the understory, and layers of litter (26); and (3) impeding the velocity and erosive ability of surface runoff by stems, surface roots, and litter. These roles are imitated by integral agroforestry. Inherently deep-rooted tree species are selected not only for their anchorage functions but also to minimize the competition of the trees with annual crops for nutrients on the surface soil and to reduce the interference of these roots with cultivation. Furthermore, in applying one of the spatial arrangements available to integral agroforestry shown earlier, trees can be arranged closely in hedgerow fashion along contours to serve as efficient erosion barriers, and the crops placed between rows (Fig. 2). The soil- and nutrient-conservation effect of this "alley cropping" type of spatial arrangement closely approximates that of terracing and is one reason why annuals could be continuously raised without drastically reducing site productivity.

14

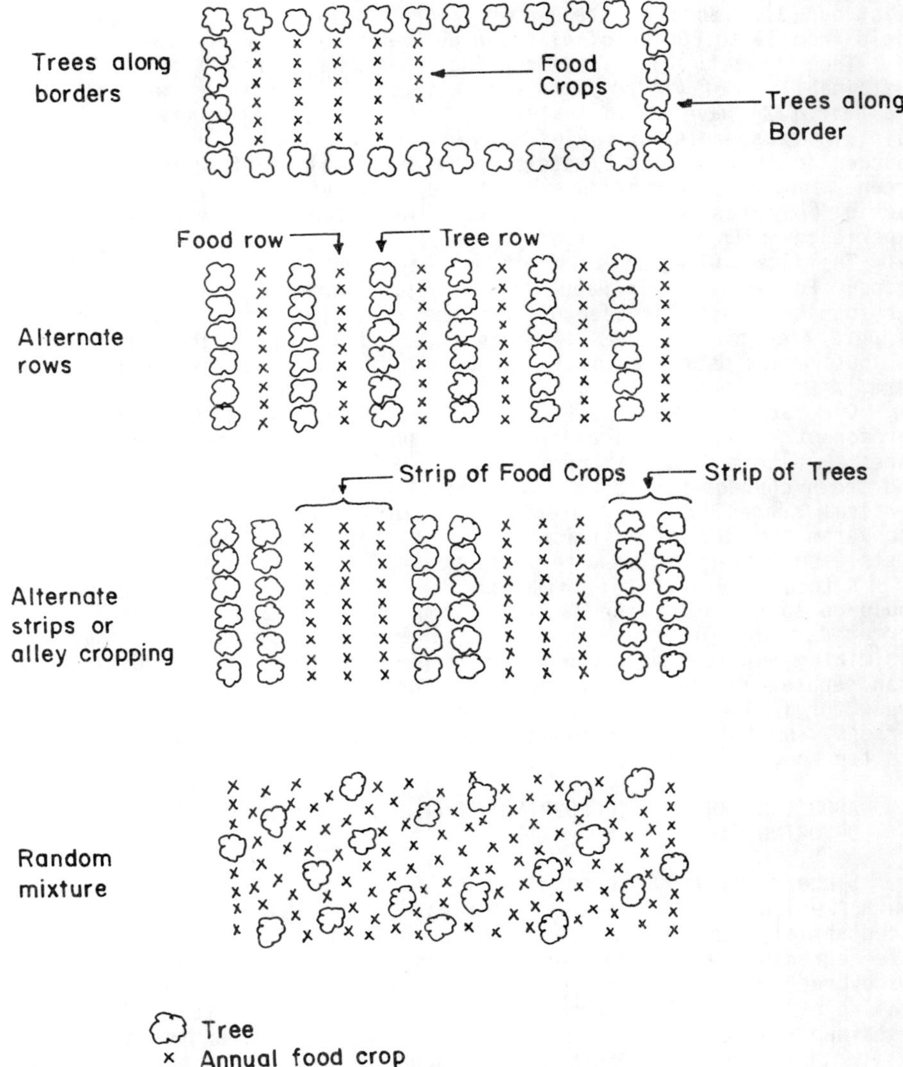

Figure 5.   Integral agroforestry system under various spatial arrangements (24).

One other important aspect of integral agroforestry, usually absent in traditional swidden, is the control over tree species composition. Over and above selecting those that function well in soil conservation, species are often chosen also for their level of contribution to the human or animal nutrient budget. For instance, nitrogen, frequently limiting in many farm situations (1), could be continuously added into the ecosystem by planting tree species capable of biological nitrogen fixation (3). Some of the better known legume trees, for instance, can yield from 80 to 600 kg of nitrogen per hectare per year (13).

The impact of intercropping N-fixing trees on levels and sustainability of yields can be dramatic. For instance, Kang et al. (9) and Nair (12) have shown that alley cropping with <u>Leucaena</u> in Africa can maintain reasonable corn yields over time. Pacardo (16) has noted that <u>Leucaena</u> contour hedges 5 meters apart can supply sufficient nitrogen-rich green manure to approximately double corn yields. In South America, Rachie (18) has shown results of alley cropping consistent with the experience in Asia and Africa.

The flow of benefits is not one way from the tree crops to the food crops. For example, in Papua New Guinea, growth of eucalypts in integral agroforestry farms has been noted to be about 10 percent higher than those of pure tree plantations. Apparently, ridding the agricultural crops of noxious weeds simultaneously benefits the tree crops (Cortes, pers. comm.).

One other reason for the potential sustainability of integral agroforestry is its imitation of the polycultural character of natural forests, albeit in a modified manner. Multiple species of various sizes and crown characteristics promote efficiency in the use of horizontal and vertical spaces, solar radiation, moisture, and other resources of the ecosystem and this diversification may further reduce the risk from forest pests (see Altieri et al., this volume).

Although both rotational and intercropping types of agroforestry could be sustainable, the latter or "modern" type appears superior to the former for the following reasons: (1) it enhances nutrient inputs while minimizing nutrient withdrawal simultaneously with crop production rather than separately; (2) it may provide continuous food production without a break for fallow; (3) it requires less land to support the same number of persons, and (4) the simultaneous presence of annuals and perennials means greater species diversity and ecological stability.

2.5. ADOPTION OF INTERCROPPING TYPE AGROFORESTRY: CONSTRAINTS AND OPPORTUNITIES

Since swidden has been in practice in the tropical parts of Asia, South America, and Africa for centuries, there seems no question about its acceptability to farmers. However, because of its reputation as a site-degrading system, and because it is rapidly losing even its "recently discovered" sustainability due to shortened fallows and reduction of available new land, there is a need to replace it with the more sustainable intercropping version of agroforestry. The immediate task is to identify the constraints to the adoption of intercropping and assess opportunities for overcoming them.

2.5.1. <u>Land Tenure</u>

Success in motivating upland farmers to practice soil conservation through integral agroforestry depends to a large extent on their perception of the economic rewards from such practice, which is greatly influenced by the nature of their tenure over land and vegetation. If they have long-term access and control over the farms, they will consider it more worthwhile to apply productivity-maintaining techniques. On the other hand, if they have insecure tenure and face the risk of eviction, they would rationally try to maximize immediate yields and not worry about the future capacity to produce. The most secure form of tenure, which is usual among lowland farms, is private individual ownership. On sloping forest lands converted to swidden farms, however, the ownership is either in the hands of the state, as is often the case in Asia, or under the customary control of clans or tribes, as in the case of many South Pacific countries. In both instances, individual control and ownership is absent, and farmers would normally have minimal motivation to apply agroforestry techniques.

Modification of state or customary tenures could overcome this disadvantage. For instance, while retaining land ownership, states can encourage agroforestry practice by granting long-term leases to interested farmers, as tested in Southeast Asia recently. The Philippines, for example, has instituted a 25-year lease system, renewable for another 25 years, for public lands devoted to agroforestry (4, 17). Thailand, in a similar move, has set up usufruct arrangements in a pilot project in the Northeast where a farmer can use the land for life if agroforestry techniques are adopted. Initial farmers' responses have been positive, and those systems are now in the process of being expanded.

In the case of customary lands, Papua New Guinea's existing clan ownership system, for instance, dictates that upon harvest of food crops by a household and reversion of the site to fallow, the land automatically goes back to tribal control. However, if integral agroforestry is practiced and trees are planted, the person gains long-term tenure over these perennial crops without necessarily gaining individual tenure over the land. This mechanism satisfies both the desire of farmers to control their long-term crops in agroforestry and the wish of clans to retain communal ownership of land resources.

2.5.2. Transfer of Technology

Adoption of any new production technique depends on how such technology is conveyed to the users, and how such prospective users perceive the benefits from adopting it (see Rocheleau, this volume). If a production system, such as integral agroforestry, is devised and presented as a radical departure from the traditional farming systems that farmers are familiar with, it could be regarded as risky and would be resisted accordingly. On the other hand, if it is shown to be merely a modification and improvement of current methods, farmers would more readily adopt it (23).

The farmer's view of comparative benefits and risks from the new land-use technique is just as crucial. A marginal farmer with little or no margin for fluctuations in output will hesitate to risk the survival of his household and will avoid potentially productive systems that are untested or fraught with risks. This reluctance could often be overcome by actual demonstration rather than by theoretical presentation of new technology (10). Minimizing risk of failures by widening the range of crop species through polyculture is a sound concept, but until it is more clearly demonstrated to farmers through pilot projects, they will continue to be wary.

2.5.3. Support Services

Opening up "markets" for new technologies is difficult enough, but filling the "demand" of a newly created market is additionally tough.

Extension campaigns are at times more successful in "selling" the new technology than expected. Often, when people adopt a new technology, critical items such as seeds of preferred species, capital to fund starting projects, farm labor to cope with the increased labor requirements or technicians to guide farmers in the new farming system may be so severely limited that they constrain wide-scale adoption.

2.5.4. Marketing

When the desired sustainability and productivity of agroforestry production systems elevate yields beyond local consumption levels, the problem of disposing surplus outputs arises. Such a problem is often aggravated by the dispersal of small agroforestry farms over remote and inaccessible terrain, and by the perishability of many food products. Some participants in government agroforestry programs have been known to withdraw when frustrated by the inability to sell their products in excess of subsistence requirements.

Multiple solutions to these problems are required, such as: (1) improvement of access through accelerated infrastructure development, (2) institutionalization of cooperative marketing organizations to facilitate product assembly at key points for easier sale to buyers, (3) selection of products that have greater storability, such as grains and nuts, and (4) selection of species that need not be simultaneously harvested at specific times of maturity and can continue to grow in volume and value when left unharvested, as in the case of root crops.

REFERENCES

1. Ahmed, S. 1982. Projected nitrogen needs in the year 2000 and alternative supply sources. Workshop Biological Nitrogen Fixation, East-West Center, Honolulu, HI.
2. Ahn, P.M. 1978. The optimum length of planned fallows. In: Soils research in agroforestry, pp 15-40. ICRAF, Nairobi, Kenya.
3. Brewbaker, J.L., Van Den Beldt, R. and K. MacDicken. 1982. Nitrogen-fixing tree resources: potentials and limitations. In: P. H. Graham and S. C. Harris (eds), BNF, Technology for Tropical Agriculture, pp 413-425. Centro Investigacion Agricultura Tropical (CIAT), Cali, Colombia,
4. Bureau of Forest Development. 1984. A primer on the integrated social forestry program. Ministry of Natural Resources, Quezon City, Philippines.
5. Christanty, L. and J. Iskandar. 1982. Soil fertility and nutrient cycling in traditional agricultural systems in West Java. Envir Policy Inst, East-West Center, Honolulu, HI.
6. Conklin, H.C. 1957. Hanunoo agriculture in the Philippines. FAO Development Paper No 12, FAO, Rome.
7. de la Cruz, R. 1982. Quantity of nitrogen contents of litterfall from forest stands in Mt. Makiling, Laguna. Univ Philippines, Los Banos, Laguna, Philippines (unpub).
8. FAO. 1957. Shifting cultivation: an appeal by FAO to governments, research centers and associations. Unasylva 2.
9. Kang, B.T., Wilson, G. F. and L. Sipkens. 1984. Alley cropping: A stable alternative to shifting cultivation. Intern Inst Trop Agric (IITA), Ibadan, Nigeria.

10. Magno, V. 1982. Extension components of smallholder tree farming in the Philippines. FAO/SIDA Seminar on Forestry Extension, FAO, Rome.

11. Myers, N. 1980. Conversion of tropical moist forests. Nat Acad Sci, Washington, D.C.

12. Nair, P.K. 1984. Soil Productivity Aspects of Agroforestry. ICRAF, Nairobi, Kenya.

13. National Academy of Sciences. 1979. Tropical legumes: resources for the future. Nat Acad Sci, Washington, DC.

14. O'Loughlin, C.L. 1974. The effect of timber removal on the stability of forest soils. J Hydrol 13:121-34.

15. O'Loughlin, C.L. and R.R. Ziemer. 1982. The importance of root strength and deterioration rates upon edaphic stability in steepland forest. In: R.H. Waring (ed), Carbon uptake and allocation: a key to management of subalpine ecosystems, pp 70-78. For Res Lab, Oregon State Univ, Corvallis, OR.

16. Pacardo, E. 1981. Water relations and nutrition of some leguminous trees. Workshop on agroforestry, Envir Policy Inst, East-West Center, Honolulu, HI.

17. Payuan, E.V. 1984. Agroforestry programs of the Bureau of Forest Development. Inter-Agency Group for Research in Agroforestry, Workshop on Agroforestry, College, Laguna, Philippines.

18. Rachie, K.O. 1981. Intercropping tree legumes with annual crops. Centro Investigacion Agricultura Tropical (CIAT), Cali, Colombia.

19. Rice, D. 1979. Upland agriculture development in the Philippines: an analysis and a report on the Kalahan programs. Upland Hydroecol Prog Ann Rep, Univ Philippines, Los Banos, College, Laguna, Philippines.

20. Sabhasri, S. 1978. Effects of forest fallow on forest production and soil. In: E.C. Chapman and S. Sabhasri (eds), Farmers in the forest, P. Kunstadter, pp 160-184. Univ Press Hawaii, Honolulu, HI.

21. Sawat, D. and S. Rouysungneru. 1977. Litter accumulation of some species in forest plantations. Royal For Dept, Bangkok, Thailand.

22. Thiesenhusen, W.C. 1974. Food and population growth. Econ Fin in Indonesia 22:201-226.

23. Vayda P., Pierce, C. J. and M. Brotokusumo. 1980. Interactions between people and forests in East Kalimantan. Impacts Sci Soc 31:179-190.

24. Vergara, N.T. (ed). 1982. New directions in agroforestry: the potential of tropical legume trees. Envir Policy Inst, East-West Center, Honolulu, HI.

25. Vergara, N.T. 1985. Agroforestry systems: a primer. Unasylva 37:22-28.

26. Wiersum, K.F. 1984. Surface erosion under various agroforestry systems. IUFRO/EWC Symp Effects of Forest Land Use on Erosion Slope Stability, Envir Policy Inst, East-West Center, Honolulu, HI.

27. Ziemer, R.R. 1981. Roots and stability of forested slopes. Intern Assoc Hyrol Sci, Pub 132, Christchurch, New Zealand.

28. Zinke, P.J., Sabhasri, S. and P. Kunstadter. 1978. Soil fertility aspects of the Lua forest fallow system of shifting cultivation. In: P. Kunstadter, E.C. Chapman and S. Sabhasri (eds), Farmers in the forest, pp 134-159. Univ Press Hawaii, Honolulu, HI.

3. SOIL PRODUCTIVITY UNDER AGROFORESTRY

P. K. R. Nair

Principal Scientist, International Council for Research in Agroforestry (ICRAF), P. O. Box 30677, Nairobi, Kenya

3.1. INTRODUCTION

In the process of conceptualization and principle formulation of agroforestry, a number of hypotheses and attributes have been advanced in the recent past. One of the most significant and oft-repeated among these is the suggested potential of agroforestry as a major practical land management alternative for maintenance of soil fertility and productivity in a variety of situations in the tropics. Most of these hypotheses have not been tested and proven by research. However, past experience and accrued knowledge on the behavior and management of tropical soils suggest that some of the agroforestry options can overcome at least a few of the drawbacks of monocultural (crop) production systems.

An account of soil productivity and management considerations under agroforestry has been given by Nair (4). Wiersum (10) and Lundgren and Nair (3) elaborated the role of agroforestry in soil conservation, taken in its broader sense to mean conservation of fertility as well as prevention of erosion; this is now being followed by a detailed ICRAF review on the topic. Young (11, 12) elaborated the potential of agroforestry as a practical means of sustaining soil fertility, and discussed the research needs in this direction, especially in relation to the research objectives of the Tropical Soil Biology and Fertility Programme (TSBF). The role of agroforestry in the context of "Land Clearing and Development in the Tropics" was discussed by Nair (6) as a part of IBSRAM (International Board of Soil Research and Management) activities. Research on soil fertility aspects of alley cropping is being undertaken vigorously by IITA (International Institute of Tropical Agriculture), Ibadan, Nigeria and other research groups.

Thus, there is now a substantial volume of scientific activity related to soil fertility and productivity aspects of agroforestry, including field experiments in various phases of implementation at different places. As a consequence the principles and concepts of agroforestry are gradually being translated into management technologies. This paper evaluates the progress and summarizes the results in this direction, especially during the past 2-3 years after the publication of the author's previous work on the topic (4).

3.2. AGROFORESTRY SYSTEMS AND PRACTICES

In order to increase the understanding of agroforestry (AF) in general, and of the various productive and protective roles it plays in existing land use systems in developing economies in particular, ICRAF has conducted a rather elaborate inventory of such systems and practices existing in different parts of the world. This global project has gathered information on a few hundreds of agroforestry systems. For example, the Agroforestry System Description Series, being serialized in _Agroforestry Systems_, describes some of the prominent and promising systems; the computerized Agroforestry Systems Register at ICRAF contains data entries on about 150 systems and/or practices (7). Most, if not all, of these system records contain information (obviously in varying degrees of detail) on the soil productivity and fertility aspects of the systems concerned.

All AF systems consist of at least two of the three major groups of AF components: trees (including shrubs), agricultural crops, and pasture/ livestock, trees being present in all AF systems. Occasionally there may be other components also, such as fish, honeybees, etc. Depending on the nature and type of components involved, AF systems can be classified as agrisilvicultural (trees + crops), silvopastoral (trees + pasture and/or animals) and agrosilvopastoral (all three types of components) (5). An overview of the examples of various AF systems in different parts of the world, derived from the earlier-mentioned global inventory, has already been presented elsewhere in this volume.

An AF system consists of one or more AF practices that are practiced extensively in a given locality or area; the system is usually described according to its biological composition and arrangement, level of technical management, or socio-economic features. An AF practice, on the other hand, denotes a specific land-management operation on a farm or other management unit, and consists of arrangements of AF components in space and/or time. Several such practices are involved in the constitution of an AF system; AF practices can also be found in land-use systems that are not agroforestry systems.

The distinctive AF practices that constitute all of the AF systems recorded in various ecoregions and locations are only few. The most common AF practices have been listed elsewhere in this volume. Obviously all these AF practices have effects and influences upon soils; however, there are only about ten or so distinctive AF technologies that have potential applicability to maintain the soil's fertility and improve its productivity. These are:

- improved fallow,
- alley cropping,
- multispecies tree gardens,
- home gardens,
- plantation crop combinations,
- trees in soil conservation and land reclamation,
- shelterbelts and windbreaks
- multipurpose trees on rangelands,
- boundary planting of trees and woody hedges, and
- woodlots for green manure, woody mulch or fodder.

In order to exploit the soil improvement potential of these AF practices and devise appropriate management options, we need to understand the ways in which a particular AF practice, the tree component of it in particular, contributes to such effects.

3.3. SUGGESTED MECHANISMS OF SOIL IMPROVEMENTS IN AGROFORESTRY

3.3.1. <u>Evidence from Existing Land Use Systems</u>

Based on the evidence collated from other existing and relatively well-researched land use systems that are related to agroforestry, Nair (4) postulated the following hypotheses about the expected soil changes under AF. In doing this, the emphasis was on those systems where trees and other woody perennials did not form a significant component of the existing land use, but where such species could be integrated without unduly modifying the production patterns of the main existing enterprises.

1. The inclusion of compatible and desirable species of woody perennials on farmland can result in a marked improvement in soil fertility. There are several possible mechanisms for this, which include:

. an increase in the organic matter content of the soil through addition of leaf litter and other plant parts;
. more efficient nutrient cycling within the system, and consequently more efficient utilization of nutrients that are either inherently present in the soil or externally applied;
. biological di-nitrogen fixation and solubilization of relatively unavailable nutrients, for example, phosphate, through the activity of mycorrhizas and phosphate-solubilizing bacteria;
. an increase in the plant cycling fraction of nutrients, with a resultant reduction in the loss of nutrients beyond the nutrient absorbing zone of the soil;
. complementary interactions between the component species of the system, resulting in a more efficient sharing of nutrient resources among the components;
. an enhanced nutrient economy because of different nutrient-absorbing zones of the root systems of the component species;
. a moderating effect of additional soil organic matter on extreme soil reactions and, consequently, improved nutrient release/ availability patterns.

2. The improvement in the organic matter status of the soil can result in an increased activity of the favorable microorganisms in the root zone. In addition to the nutrient relations mentioned earlier, such microorganisms may also produce growth-promoting substances through desirable interaction and cause commensalistic effects on the growth of plant species.

3. The inclusion of trees and woody perennials on farmlands can, in the long run, result in marked improvements in the physical conditions of the soil - in permeability, water-holding capacity, aggregate stability, and soil temperature regimes. Although these improvements may be slow, their net effect is a better soil medium for plant growth.

4. The role of trees in soil conservation and erosion control is one of the most widely acclaimed and compelling reasons for including trees on farmlands that are prone to erosion hazards. The

beneficial effects of trees in this regard extend beyond protecting the immediate farmland under consideration, to imparting stability to the ecosystem and reducing the rate of siltation of down-stream aquatic ecosystems, dams and reservoirs.

5. The influences of trees on hydrological characteristics can extend from the micro-site to the farm and regional levels. Although the effect of water use by a tree component on water availability to crop plants in different climatic conditions is not yet fully understood, there is evidence that the hydrological characteristics of catchment areas are favorably influenced by the presence of trees.

3.3.2. Effect of Trees on Soils

Sanchez et al. (8) examined the role of trees as tropical soil improvers primarily from the points of view of plantation forestry. They surmised that some of the beneficial effects of trees on soils after the trees close canopy are unexpected, particularly the influence on chemical properties, such as Ca accumulation by _Gmelina arborea_, and that the main deleterious effects of tree crops on soil properties occur during their establishment phase. Discussing this topic, Young (12) identified several positive and a few negative aspects of trees on soil, some well proven and others reasonable suppositions. These, as well as the above-mentioned hypotheses of Nair (4), are summarized in Table 1, which indicate the mechanisms by which the generally favorable effects of trees upon soils may be achieved under appropriate conditions. Table 1 should not be interpreted to mean that trees have only favorable effects on soils. Some adverse effects have also been reported, notable among which include:

- fast growing species (of trees) may place a heavy demand on soil moisture, and unless properly managed, this can lead to adverse effects, especially in dryer environments;
- nutrient losses from whole-tree harvesting may be excessive, especially in plantation forestry;
- nutrient depletion may cause temporary deprival of nutrients to adjacently growing, less competitive species (although litter fall, root biomass, etc., may compensate for this to some extent in the long run);
- a badly-managed tree stand, as with any vegetation, can cause accelerated erosion;
- there may result adverse chemical/biological effects from certain tree species, leading to: acidification, allelopathy, accumulation of toxic exudates, the provision of alternate hosts of pests and pathogens, etc.; and
- shading and changes in spectral quality of light on the growth of other species in close proximity (though not a direct effect on the soil, this can have considerable significance in agroforestry).

Most of these attributes (both beneficial and adverse) are either inferred from land-use systems related to agroforestry, or are still only untested hypotheses. The magnitude of benefit or adverse effect that could be experienced will depend upon a number of site-specific factors. Moreover, many of the attributes of trees, as compared to annual crops, can be realized only over a relatively longer period of time. One of the main issues to be considered here is the soil fertility improvement in agroforestry through addition of plant biomass vis-a-vis other uses for the biomass (fodder, fuelwood, food, etc.). Similarly the proportion of space occupied by tree components in an AF combination, in order to derive the suggested benefits to any tangible extent will depend upon several factors such as soil type, plant (tree) species, tree-crop arrangement, etc. All these factors need to be taken into account while suggesting plant and soil management technologies in agroforestry.

3.4. SOIL MANAGEMENT CONSIDERATIONS

The favorable effects (in Table 1), in so far as they are operative, indicate that the incorporation of soil-improving and compatible woody species on farmlands can considerably retard the rates of undesirable processes of soil degradation and productivity decline that are experienced in many tropical farmlands by current farming practices. However, our knowledge of soil conditions in agroforestry is much less than that in well researched agricultural and forestry land-use systems. Validation of the aforementioned hypotheses through appropriate research is essential before the practices based on them can be recommended for adoption with any reasonable assurance of benefits. Nevertheless, in order to circumvent the considerable time-lag in generating research data from well conducted experiments, it seems justifiable to consider some expedient measures, however ad hoc these might appear to be at present. Some of the recent results from alley cropping experiments at IITA (2, Table 2), illustrate the feasibility of this technology (and other forms of hedgerow intercroppings (9)). ICRAF is currently preparing a technology package on alley cropping, which will contain various management considerations on this technology for different ecological regions.

Similarly, the evidence of the soil-improving qualities of some multipurpose tree species (e.g., Acacia albida and Prosopis cineraria) in the dry regions, and their complementarity with the understorey agricultural species would seem to confirm the distinct promise held out by these species. Therefore, one may expect improved soil management systems for other marginal areas and "wasted" lands in different ecological situations involving a wide range of useful woody perennial species. ICRAF, through its large project on "Multipurpose Tree (MPT) Database" has assembled a substantial body of information on multipurpose trees with potential role, in various ways, in agroforestry under different conditions (1). The earlier- mentioned AF System Description Series and other outputs from the Agroforestry Systems Inventory Project in ICRAF (7) also provide information on the soil enriching attributes of several species of multipurpose trees and shrubs in a large number of indigenous systems.

In these circumstances, some approaches to devise soil management technologies for low-input agroforestry systems can be suggested:

1. Certain trees and other woody perennials can be incorporated on farmlands preferably without causing significant changes in conventional agricultural practices. Examples include alley planting, other zonal systems, and contour strips. Similarly, the intercropping of agricultural species can be undertaken in tree stands in a number of ways. Several approaches have been

Table 1. Summary of the potential beneficial effects of trees on soils

Nature of Processes	Processes	Main Effect on Soil	Scientific evidence
Input processes (augment additions to the soil)	Biomass Production	Addition of carbon and its transformations	Available
	Nitrogen fixation	N-enrichment	Available
	Rainfall	Effect on rainfall (quantity and distribution) and therefore nutrient addition through rain.	Not adequately demonstrated
Output process (reduce losses from the soil)	Protection against water and wind erosion	Reduce loss of soil as well as nutrients	Available
	Increase evapotranspiration	Reduce runoff, concentrate nutrients	Available
Turn-over processes	Nutrient retrieval/cycling/release	Uptake from deeper layers and "deposition" on surface via litter	Not adequately demonstrated
		Witholding nutrients that can be lost by leaching	Not demonstrated
		Timing of nutrient release: this can be regulated by management interventions	Available

Table 1.   (Continued)

"Catalytic" processes (indirect influences)	Physical processes	Improvement of physical properties (water-holding capacity, permeability, drainage, etc.) at the microsite as well as at the watershed (macrosite) level	Available
	Root growth and proliferation (enhanced)	Addition of (more) root biomass; growth promoting substances; micrbial associations	Partially demonstrated
	Litter quality and dynamics	Improvement of litter quality through diversity of plant species; better timing of quantity, and method of application of litter possible	Now being increasingly studied in alley cropping and other imtercropping experiments
	Microclimatic processes	Creation of more favourable microclimate; shelterbelt and windbreak effects	Available
	(Bio)chemical/biological processes (net effects of various processes)	Moderating effect on extreme conditions of soil acidity, alkalinity, etc.	Partially demonstrated

Table 2. Some results of alley cropping experiments at IITA, Ibadan.
Nigeria (Source: Ref. 2).

A: The effect of <u>Leucaena</u> prunings and nitrogen on the grain yield (tons
ha^{-1}) of maize variety TZPB alley cropped with <u>L. leucocephala</u>.
Unfortunately the maize crop was seriously affected by drought during
early growth.

N fertiliza-tion rate ($kg\ N\ ha^{-1}$)	Leucaena prunings	Year 1979	1980	1981	1982	1983
0	Removed	-	1.04	0.48	0.61	0.26
0	Retained	2.09	1.91	1.21	2.10	1.92
80	Retained	3.54	3.26	1.89	2.91	3.16
LSD 0.05		0.36	0.31	0.29	0.44	0.79

B: Chemical properties of surface soil (sandy Entisol) after six years
of alley cropping <u>Leucaena leucocephala</u> with maize and cowpeas.

Treatments (kg N/ha)	Leucaena prunings	pH-H_2O	Org. C. (%)	K	Exchangeable Ca (me/100g)	Mg	P (ppm) (Bray No. 1)
0	Removed	6.0	0.65	0.19	2.90	0.35	27.0
0	Retained	6.0	1.07	1.28	3.45	0.50	26.2
80	Retained	5.8	1.19	0.26	2.80	0.45	25.6
LSD 0.05		0.2	0.14	0.05	0.55	0.11	5.3

proposed to increase crop production in shifting cultivation
systems without substantially changing the structure of such
systems (e.g., the corridor system, shortening of the fallow
period, planted fallows and other ways of improving the quality of
fallows).

2. Methods of land clearing and site preparation are of crucial
   importance because certain mechanical operations can result in
   serious damage to soil physical properties, leading to compaction
   and degradation of soil structure or the loss of topsoil by
   erosion. Similarly, conventional land preparation methods,
   especially for agricultural species, can aggravate soil erosion
   and impair soil physical conditions. The choice of land clearing
   and preparation methods depends on soil properties, species, and
   anticipated intensity of management. The magnitude of effort

needed to control weeds decreases as the proportion of the soil surface that is left unprotected by a useful plant canopy over a specified time span decreases.

3. When relatively short-duration agricultural species are continuously cultivated in pure stands or in combination with perennial species, the fertility status of the soil will change, necessitating frequent external inputs of nutrients as manures and fertilizers in order to compensate for frequent "export" of nutrients from the soil through harvests. However, in many areas the cost and/or availability of fertilizers make heavy fertilization uneconomical and impractical. It is in this context that it becomes important to exploit the desirable soil-enriching and/or restoring characteristics of woody perennials to the fullest extent.

4. A combined stand of plants of different growth habits and forms can be of considerable advantage in protecting the soil also. The presence of a greater plant cover over soil, either as living plants or as mulch, reduces the impact of raindrops on the soil surface and thus minimizes splash and sheet erosion. Moreover, a higher organic matter content and a larger root volume in the soil impart better physical conditions, thus causing increased infiltration and decreased runoff.

Considering the vastness and complexity of situations under which agroforestry is practiced, it is inevitable that the soil management considerations outlined above have to be of a general nature. The specific management practices for a given set of conditions will depend upon the prevailing soil conditions, climate, plant species, level of management and other local situations. It would therefore appear worthwhile to aim at a soil quality categorization for grouping soils and soil conditions according to the nature of problems they present, and suggest agroforestry practices for the management of their physical and chemical properties. While the different categories so developed could indicate the main soil-related constraints, it might be possible to examine ways of overcoming such constraints by appropriate agroforestry technologies. When the necessary data become available, this scheme of "matching soil conditions vs. agroforestry technologies" would be directly applicable to land evaluation exercises, and could serve as a useful tool for soil constraint analysis vis-a-vis agroforestry solutions and alternatives.

3.5. CONCLUSIONS

High expectations have been raised in recent years about the potential of agroforestry as a major land management alternative for maintenance of soil fertility in the tropics. These are based on the relatively satisfactory situation with respect to soil conditions in several prominent and promising agroforestry systems and the postulations on the favorable effects of trees on soils. Recent research on the soil fertility aspects of alley cropping, a comprehensive review being undertaken at ICRAF on the role of agroforestry in soil conservation, an operational computer model to predict soil changes under agroforestry (11), assessment of the potential role of multipurpose trees in various aspects of agroforestry, etc., have contributed to building up a substantial body of largely descriptive knowledge related to this topic.

What is now required is a two-pronged approach: (a) validation through field testing and research, of the various hypotheses and postulations, in a coordinated, systematic manner; and (b) application of available knowledge through good land-use planning and management.

Both approaches should proceed simultaneously and iteratively. In time, our understanding of soil productivity and management under agroforestry will be much better than it is today; this will enable us not only to improve the science and practice of agroforestry, but also to eliminate some of the myths and exaggerated claims being advanced on its virtues.

REFERENCES*

*   Only some of the recent publications on the topic that have appeared after the preparation of the manuscript of (4) have been referred to in the text; readers are directed to publication No. 4 cited below for the earlier bibliography.

1.   Carlowitz, P. V. 1986. Computerized multipurpose tree (MPT) database. ICRAF, Nairobi, Kenya.
2.   Kang, B. T., Wilson, G. F. and T. L. Lawson. 1985. Alley cropping: a stable alternative to shifting cultivation. Intern Inst Trop Agric (IITA), Ibadan, Nigeria.
3.   Lundgren, B. O. and P. K. R. Nair. 1985. Agroforestry for soil conservation. In: S. A. El-Swaify, W. C. Moldenhauer and A. Lo (eds), Soil erosion and conservation, pp. 703-717. Soil Conserv Soc Amer, Ankeny, IA. (Reprinted as ICRAF Reprint 21, 1985).
4.   Nair, P. K. R. 1984. Soil productivity aspects of agroforestry. ICRAF, Nairobi, Kenya.
5.   Nair, P. K. R. 1985a. Classification of agroforestry systems. Agrofor Syst 3:97-128 (Reprinted as ICRAF Reprint 23, 1985).
6.   Nair, P. K. R. 1985b. Agroforestry in the context of land clearing and development. Working Paper No 33, ICRAF, Nairobi, Kenya.
7.   Oduol, P. A. and P. K. R. Nair. 1986. Computerized agroforestry systems register (Database). ICRAF, Nairobi, Kenya.
8.   Sanchez, P. A., Palm, C. A., Davey, C. B., Szott, L. T. and C. E. Russell. 1985. Tree crops as soil improvers in the humid tropics? In: M. G. R. Cannell and J. E. Jackson (eds), Attributes of trees as crop plants, pp. 327-358. Inst Terr Ecol, Huntington, England.
9.   Ssekabembe, C. 1985. Perspectives on hedgerow intercropping. Agrofor Syst 3:339-356.
10. Wiersum, K. F. 1984. Surface erosion under various tropical agroforestry systems. In: C. L. O'Laughlin and A. J. Price (eds), Symposium on effects of forest land use on erosion and slope stability, pp. 231-240. East-West Center, Honolulu, HI.
11. Young, A. 1985. The potential of agroforestry as a practical means of sustaining soil fertility. Working Paper 34, ICRAF, Nairobi, Kenya.
12. Young, A. 1986. Effects of trees on soils. Working Paper, ICRAF, Nairobi, Kenya.

4. SIGNIFICANT NITROGEN FIXING TREES IN AGROFORESTRY SYSTEMS

James L. Brewbaker

President, Nitrogen Fixing Tree Association and Professor of Horticulture and Genetics, University of Hawaii, 3190 Maile Way, Honolulu, Hawaii 96822

4.1. ABSTRACT

Nitrogen fixation characterizes about 650 known tree species and several thousand suspected ones. Most of these N-fixing trees (NFT) are rhizobially-nodulated legumes, largely tropical or subtropical in origin. At least 9 other plant families are implicated through actinomycete associations. Many N-fixing species are shrubs or small trees of secondary forests and grasslands. They often lend themselves better to crop- and animal-based agroforestry systems than do the premier forest timber species.

Characteristics are summarized here for 85 NFT species or species complexes that are potentially important in agroforestry systems. High priority trees are summarized for their uses as fodder, green manure, fuelwood, pulp, timber, shade and windbreak. N-fixing trees assume special importance in agroforestry systems as a source of high-nitrogen green manure, resulting in significant soil amelioration.

Many NFT species are both multipurpose and fast-growing. Site-specific, short-duration comparisons of different N-fixing tree species in small plots should be considered a matter of urgency in much of the tropics. They constitute a primary tool for counteracting the awesome losses of tropical deforestation.

4.2. INTRODUCTION

N-fixation characterizes most legumes (over 90% of mimosoids and papilionoids, and 34% of caesalpinioids). At least 90% of these represent tropical centers of origin. Selected genera in 9 other plant families also fix nitrogen: Betulaceae, Casuarinaceae, Coriariaceae, Cycadaceae, Elaeagnaceae, Myricaceae, Rhamnaceae, Rosaceae and Ulmaceae. Legume N-fixing nodules are rhizobially infected, while those of the other families involve actinomycetes of the genus _Frankia_. Nitrogen fixation characterizes about 650 known tree species and several thousand suspected ones (1, 9).

The list of potential NFT species for use in agroforestry is expanded greatly if one generously includes all plants from which wood is an economic product. Many N-fixing species are shrubs or secondary forests and grasslands. They often lend themselves better to crop- and animal-based agroforestry systems than the premier forest trees do. We cannot be so myopic as to let our classical definition of "tree" truncate the list of wood-producing perennials useful in agroforestry systems. Forestry tradition might restrict "agroforestry" to the inclusion of trees with single stems exceeding 3 meters in height, but this is purely academic to the small farmer.

N-fixing trees and shrubs that have been used in agroforestry systems or appear to have promise for such use are described briefly here (Table 1). Boland (2) provides useful Australian examples of the fuelwood, roundwood, fodder, fence, shade, windbreak, erosion control, and soil improvement categories that are intrinsic to the definition of "multipurpose" (MPT) trees. N-fixing species often provide convenient examples of multipurpose trees (7), i.e., those maintained or introduced into agroforestry systems specifically for one or more of the purposes cited by Boland.

Useful general references to these 85 species and species complexes include the U. S. National Academy of Science publications (14, 15) on legumes, acacias, calliandra, casuarina and leucaena; publications of the Nitrogen Fixing Tree Association (NFTA), including the NFTA HIGHLIGHTS and the journals "Leucaena Research Reports" and "Nitrogen Fixing Tree Research Reports"; and publications by Brewbaker (3, 4), Burley and Von Carlowitz (7), Le Houerou (11), Panday (15), Robinson (16), and Singh (17). Turnbull (18) and colleagues have illustrated the significance of N-fixation among lesser known trees with agroforestry potential in Australia. Of 100 species they describe, 63 are known or presumed to fix nitrogen. A high proportion of shrub and tree species noted as significant for energy production by the U. S. National Academy of Science (14) were also N-fixing.

4.3. FODDER TREES

N-fixing trees assume a distinct role in agroforestry systems almost entirely due to nitrogen itself. Leaf protein values of tropical and temperate legumes average about twice those of the grasses (16% to 9%). Superior leguminous fodder shrubs can exceed 20% crude protein (expressed as %N x 6.25) in animal-browse fractions (12). Fodder and green manure uses of N-fixing trees are therefore those of major importance in agroforestry. Many leguminous NFT species have high fodder values (15, 16, 17). About 80 NFT species each are viewed as significant fodders for Asia and Africa (3, 4). It must be stressed that fodder from these woody legumes is significant primarily as a supplement to high-energy tropical fodders, usually grasses. These high-energy fodders, however, are so low in protein that this can become a primary limiting factor for animal growth and reproduction (11, 12).

Among the most significant fodder shrubs and trees are legumes of the genera _Chamaecytisus_, _Desmanthus_, _Desmodium_, _Gliricidia_, _Leucaena_, _Prosopis_ and _Sesbania_, with one or two species of the giant genus _Acacia_.

4.4. ALLEY FARMING AND NURSE TREES

Nitrogen is the primary limiting nutrient in many tropical soils for crop and animal production. N-fixing trees play major roles in alley farming systems by providing green manure to intercropped food plants (10). They serve also as nurse trees in grass pastures or among food or industrial trees in plantations, and provide green vegetation which can be used directly as fertilizer or compost. Tropical soils have high rates of N turnover, and N is stored primarily in the biomass rather than in the soil. Typical values of temperate forests show 90% of the ecosystem N to be in the soil, while it is common to find less than 10% in the soil in the tropics. Clear-cut deforestation and most traditional agricultural practices in the tropics thus can leave the system grossly N-deficient.

As an example, few agricultural soils in the tropics have N levels adequate to support 3 successive maize crops yielding 2 tons/hectare of grain each. As a result, studies with leucaena leaves applied to maize as a green manure have shown excellent and linear responses of grain yield (8, 10).

Soil amelioration and green manure uses of woody legumes are most effective through alley farming systems with some alteration of rows of crop and woody hedge (10). Significant legume trees for these systems must be easily coppiced and rapidly growing, and are more farm-acceptable if they are also good fodders. They include species of the genera Calliandra, Flemingia, Gliricidia and Leucaena.

The nurse-tree role of N-fixing trees in plantations is also important. High-nitrogen leaf litter may in fact be more important to a crop than the shade or wind protection provided. Significant nurse-tree use has been made of species of Erythrina, Gliricidia, Inga, Leucaena, Mimosa, Robinia and Sesbania.

In natural ecosystems, it is common to find species paired either in space or time where one is a nitrogen fixer (e.g., eucalypts and acacias, pines and alders). It is the rare tropical forest that is not provided with N-fixing woody species in numbers adequate to ensure a continuing input of N. Inputs can be directly to the soil through root and nodule turnover, or indirectly through uptake and return via litterfall. By virtue of their small leaflets, most mimosoids provide leaf litter that decomposes and recycles nutrients rapidly (8). Many N-fixing trees are also aggressive invaders of disturbed sites. Although often unattractive, they can restore some health to soils that are virtually sterilized by removals, fire, or through invasion by perennial grasses that often dominate deforested sites. Not all legumes improve soils or control erosion simply because they fix nitrogen, to be sure, although the majority can be considered soil ameliorating plants.

4.5. FUELWOOD AND CHARCOAL TREES

Legume trees provide some of the premier woods for charcoal and firewood, of which the mesquites or algarobas (Prosopis spp.) are best known. Despite fast growth, leguminous hardwoods often have high wood density and low moisture values, allowing harvest as excellent fuelwood in 3-5 years (5, 6, 14). Examples include the genera Acacia, Acrocarpus, Calliandra, Casuarina, Gliricidia, Leucaena, Mimosa, Robinia and Samanea.

Calorific values of fresh wood (40-60% moisture) are very similar among these trees (averaging 19500 MJ/kg for dry wood) and are largely related to moisture contents. They stabilize early during tree growth. Increasing use is made in the tropics of such trees in systems to generate electricity, to generate producer gas ("gasification"), or to produce charcoal. Significant research progress has also been made in converting such biomass to liquid fuels.

4.6. FOOD AND MEDICINAL TREES

The adoption of trees in agroforestry is more commonplace if significant foods, gums, or medicinals can also be harvested. The legumes are notable for gums, especially from the genus Acacia. Legume pods and seeds often serve as food, and their seeds, leaves or barks as medicinals. Examples include species of the genera Cajanus, Hippophaea, Inga, Leucaena, Parkia, Pongamia, Sesbania and Tamarindus.

4.7. WINDBREAK AND FENCE ROW TREES

One of the most common agroforestry systems is that of trees in windbreak or fence rows near crops or animal compounds (see also Budowski, this volume). N-fixing trees that excel for this use often are easily propagated vegetatively. Among the genera most used are Baphia, Cajanus, Calliandra, Casuarina, Elaeagnus, Erythrina, Gliricidia and Pithecellobium.

Many woody legumes are thorny, conferring major protection against herbivores. They are occasionally planted or lopped to make thorn hedges that restrain feral animal movement. Despite these traditional uses, it is not generally acceptable to recommend introduction of thorny species as exotics. Indeed, careful selection is underway to produce thornless populations of species such as Prosopis alba and P. pallida/juliflora.

4.8. PULPWOOD AND ROUNDWOOD TREES

It may seem unfair to assign an apparently low priority to the traditional commercial uses of trees (timber, poles, pulp) by listing them last. Although agroforestry systems involving animals with such commercial species do exist (see, for example, Lewis and Pearson, this volume), they are normally best considered as species for solid stand (plantation) management. Major concerns have surfaced about the damage caused to neighboring crops by tall trees, notably eucalypts, largely from experiences with fence row plantings. Among the causes of crop yield reductions cited, are clearly the reduction of incident light to the crop, of greatest importance to C4 cereals, and the reduction of available water. Allelopathy of fallen leaves and reduction in available nitrogen (from non-NFT species) are other cited causes. Obviously, the degree of acceptable crop yield losses depends on the objectives of the particular agroforester.

In addition to fuel, small poles are the major important wood product of trees from agroforestry systems, and many NFT genera provide useful polewood and often pulpwood or craftwood (Table 1). These include species of the genera Acacia, Albizia, Alnus, Baphia, Casuarina, Dalbergia, Inga, Leucaena, Mimosa, Ougeinia, Parkia, Periserianthes, Pterocarpus and Samanea.

REFERENCES

1. Allen, O. N. and E. K. Allen. 1981. The Leguminosae: a source book of characteristics, uses, and nodulation. Univ Wisc Press, Madison, WI. 812 p.
2. Boland, D. J. 1986. Seed nursery practice and establishment. In: J. W. Turnbull (ed), Multipurpose Australian trees and shrub, pp 59-80. Australian Council Intern Agric Res, Canberra.
3. Brewbaker, J. L. 1985. Leguminous trees and shrubs for Southeast Asia and the South Pacific. In: G. J. Blair (ed), Forages in Southeast Asian and South Pacific agriculture, pp 43-50. Australian Council Intern Agric Research, Canberra.
4. Brewbaker, J. L. 1986. Nitrogen fixing trees for fodder and browse in Africa. In: B. T. Kang (ed), Alley farming for humid and subhumid regions of tropical Africa. Intern Inst Trop Agric (IITA), Ibadan, Nigeria.

5.   Brewbaker, J. L., Van Den Beldt, R. and K. MacDicken. 1982.
     Nitrogen-fixing tree resources: potentials and limitations.
     In: P. H. Graham and S. C. Harris (eds), BNF Technology for Tropical
     Agriculture, pp 413-425. Centro Investigacion Agricultura Tropical
     (CIAT), Cali, Columbia.
6.   Brewbaker, J. L. Van Den Beldt, R. and K. MacDicken. 1984. Fuelwood
     uses and properties of nitrogen-fixing trees. Pesquisa agropec bras
     19:193-204.
7.   Burley, J. and P. Von Carlowitz (eds). 1984. Multipurpose tree
     germplasm. Intern Council Res Agrofor (ICRAF), Nairobi, Kenya.
8.   Evensen, C. L. I. 1984. Seasonal yield variation, green leaf
     manuring, and eradication of _Leucaena leucocephala_. MS Thesis, Univ
     Hawaii, Honolulu, HI.
9.   Halliday, J. 1984. Register of nodulation reports for leguminous
     trees and other arboreal genera with nitrogen fixing members.
     Nitrogen Fixing Tree Res Rep 2:38-45.
10. Kang, B. T., Wilson, G. F., and T. L. Lawson. 1986. Alley cropping: A
     stable alternative to shifting cultivation. Intern Inst Trop Agric
     (IITA), Ibadan, Nigeria. 22 p.
11. Le Houerou, H. N. (ed). 1980. Browse in Africa. Intern Livestock
     Centre Africa (ILCA), Addis Ababa, Ethiopia.
12. Minson, D. J. and J. R. Wilson. 1980. Comparative digestibility of
     tropical and temperate forage--a contrast between grasses and
     legumes. J Austr Inst Agric Sci 1980:247-249.
13. NAS. 1979. Tropical legumes: resources for the future. US Nat Acad
     Sci, Washington, D.C. 331 p.
14. NAS. 1985. Firewood crops: Shrub and tree species for energy
     production (2 Vols). US Nat Acad Sci, Washington, D.C.
15. Panday, K. 1982. Fodder trees and tree fodder in Nepal. Swiss Fed
     Inst For Res, Birmensdorf, Switzerland. 107 pp.
16. Robinson, P. J. 1985. Trees as fodder crops. In: M. G. R. Cannell and
     J. E. Jackson (eds), Attributes of trees as crop plants, pp 281-300.
     Inst Terrestrial Ecol, Huntingdon, UK.
17. Singh, R. V. 1982. Fodder trees of India. Oxford and IBH Publ Co, 66
     Janpath, New Delhi, India. 663 pp.
18. Turnbull, J. W. (ed). 1986. Multipurpose Australian trees and shrubs.
     Australian Council Intern Agric Res, Canberra. 316 pp.

Table 1. Significant N-fixing trees and shrubs in agroforestry systems.
Descriptions follow this format:

SPECIES EPITHET, "Common Names", (SUBFAMILY; FAMILY) (of first species
    listed in genus)
    1.   Center of Origin; Distribution; Uses
    2.   Description; Botany; Ecology

ACACIA ALBIDA Del. "Winterthorn", "Kad" (MIMOSOIDEAE; LEGUMINOSAE)
    1.   Africa; widespread, now to India, Israel; fodder, ornamental,
         shade, green manure
    2.   Tree to 20 m; = A. leucophloea; leafless in rainy season;
         bipinnate; thorny; drought tol (tolerant)(to 300 mm); frost sens
         (sensitive)

ACACIA ANEURA F. Muell. ex Benth. "Mulga"
    1.   Australia, widespread (150 mha); hard wood, fuelwood,
         ornamental, variably browsed by stock, with DMD (dry matter
         digestibility) a low 39%
    2.   Shrub or tree to 12 m; slow growth; phyllodes; high drought tol
         (to 200 mm); wide range of frost tol

ACACIA ANGUSTISSIMA Miller
    1.   Central America; now to Indonesia; fuelwood, fodder (poor)
    2.   Shrub to 5 m; rapid growth; not thorny

ACACIA AULACOCARPA A. Cunn. ex Benth. "Brown salwood"
    1.   N and NE Australia; timber, fuelwood, shade
    2.   Tree to 30 m, or variably shrubby; tropic rainforests

ACACIA AURICULIFORMIS A. Cunn. ex Benth. "Northern black wattle"
    1.   N.E. Australia, Papua New Guinea; now widely in S Asia;
         fuelwood, pulpwood, shade, ornamental, erosion control
    2.   Irregular tree to 30 m, usually less, branched; phyllodes; tol
         of waterlogged, saline or acid soils; humid tropics

ACACIA CONFUSA Merr. "Formosan koa"
    1.   Taiwan, Philippines; firewood, ornamental, soil amelioration
    2.   Spreading tree to 14 m; mesic subtropics (to 750 mm), acid tol

ACACIA CORIACEA DC "Wirewood", "Desert Oak", "Dogwood"
    1.   N Australia, 1.5 mha; introduced in Africa; fodder (pods,
         leaves) of poor palatability, fuelwood, fenceposts, shelter
    2.   Shrub to 7 m; phyllodes; slow growth; high tol of drought and
         fire.

ACACIA CRASSICARPA A. Cunn. ex Benth. "Northern wattle"
    1.   NE Australia, Papua New Guinea; timber, flooring, dune
         stabilization, shade
    2.   Tree to 20 m, variable; phyllodes; fast-growing; humid tropics,
         coastal; acid tol

ACACIA FARNESIANA (L.) Willd. "Cassie", "Huisache" (Mex), "Sweet Acacia",
    "Mimosa Bush", "Klu" (Hawaii)
    1.   Americas, now worldwide; ornamental, cult for perfume, tannin,
          dyes, gums; fodder cyanogenic, young pods browsed
    2.   Shrub 2-7 m; bipinnate; forms thickets; stipular spines (<1 cm);
          frost tol; rapid growth; weeds; tol of heavy clay

ACACIA HOLOSERICEA A. Cunn. ex G. Don
    1.   N. Australia; grows well in Africa; fuelwood, ornamental,
          hedges, fodder (only fed dry)
    2.   Shrub to 6 m; phyllodes; subhumid tropics; widely adaptable, tol
          of salinity and to short periods of drought

ACACIA LITAKUNENSIS Burch. "Umbrella thorn"
    1.   Africa (South); fodder (pods)
    2.   Tree to 10 m; thorny; bipinnate; forms thickets

ACACIA MANGIUM Willd. "Mangium", "Brown salwood"
    1.   NE Australia, Papua New Guinea, Indonesia; timber, pulp,
          fuelwood
    2.   Stately tree to 30 m; large phyllodes; humid tropics; acid soil
          tol

ACACIA MEARNSII de Wild. "Black wattle"
    1.   SE Australia; now panboreal; tannin, paper pulp, charcoal,
          fuelwood, poles, erosion control, shelterbelt, firebelt,
          ornamental; fodder acceptable only in mixtures
    2.   Large shrub or small tree to 18 m; aggressive growth, even
          weedy; moist sub-tropics, mid-elevations, mild frost tol

ACACIA NERIIFOLIA A. Cunn. ex Benth. "Silver Wattle"
    1.   E Australia; windbreaks, emergency stock fodder, possible
          fuelwood, tannin, ornamental
    2.   Shrub to 8 m or tree to 12 m; phyllodes; drought tol (to 300 mm)
          and frost tol, fast-growing

ACACIA NILOTICA (L.) Willd. ex Del. "Babul" (India), "Munga" (Africa)
    "Prickly Acacia" (Australia)
    1.   Africa, India; now widespread; firewood, charcoal, tannin and
          gum source, fodder (pods, leaves)
    2.   Tree 6-12 m; very thorny; bipinnate; wide range in tropics to
          midlands and highlands; frost sens, highly drought tol,
          deciduous

ACACIA PENDULA A. Cunn ex G. Don. "Myall"
    1.   E Australia; widely introd. in Near East; drought-stock fodder
          (DMD 47%), shade and windbreak tree, timber, fuelwood,
          ornamental
    2.   Tree to 8 m; stately; phyllodes; subtropical drought tol to 200
          mm

ACACIA PODALYRIIFOLIA A. Cunn. ex Benth.
    1.   NE Australia; to Africa; ornamental, bee pasture, fuelwood,
          tannin, windbreak
    2.   Shrub to 5 m, phyllodes; rapid growth, tropical, moist forests

ACACIA POLYACANTHA Willd. "Khair", "Catechu Tree"
   1.   Africa, India; good fodder (DMD 61%), charcoal, black gum, dye
   2.   Tree to 25 m; coppices well, long-lived; bipinnate; = A.
      catechu; recurved spines; weedy; midlands to 1000 m, survives
      mild frost, low drought tol

ACACIA SALICINA Lindl. "Cooba", "Willow wattle"
   1.   Australia; shade, shelters, windbreak, timber, fodder (pods,
      leaves dried)
   2.   Tall shrub or tree to 20 m; vigorous suckers, subhumid
      subtropics; drought and frost tol

ACACIA SALIGNA (Lindl.) H. Wendl. "Coolan", "Willow"
   1.   W Australia (now widespread); common in S and N Africa;
      ornamental, low-grade fodder, fuel, erosion control, dune
      stabilization
   2.   Tree to 8 m; rapid growth; phyllodes, tol of drought, fire,
      salt, wind; = A. cyanophylla

ACACIA SENEGAL (L.) Willd. "Gum-arabic Tree", "Hashab"
   1.   Africa to India; gum arabic, firewood, charcoal, fodder, soil
      amelioration, firewood
   2.   Tree to 10 m but often shrubby; deciduous; extremely thorny;
      bipinnate; dry tropics to 200 mm rain

ACACIA SEYAL Del. "Thirsty Thorn", "Dushe" (Nigeria)
   1.   Africa, N; wood, gums and tannins, important feed (pods, leaves,
      bark)
   2.   Slender tree to 12 m; bipinnate; long thorns; semi-arid tropics

ACACIA SIEBERIANA DC.
   1.   W Africa; gum, drought-season fodder, honey, furniture wood
   2.   Tree to 10 m; drought tol, but found along streams

ACACIA TORTILIS (Forsk.) Hayne.
   1.   Africa; introduced to Near East and tropical Asia; dense
      firewood, fodder (pods, leaves), soil amelioration
   2.   Flat-topped tree to 15 m, but often shrubby; thorny, bipinnate;
      dry tropics, to 100 mm; alkali tol; not frost-tol, to 1200 m in
      Kenya

ACACIA VICTORIAE Benth. "Gundabluey", "Acacia Bush", "Elegant Wattle"
   1.   Australia (widespread); fodder (pods), ornamental, windbreak
   2.   Shrub to 5 m, highly variable; straggly, thorny, fast-growing,
      short-lived, often in thickets; diverse soils; saline and
      drought tol (to 350 mm).

ACACIA VILLOSA Willd.
   1.   Caribbean, common in Indonesia; hedge, fodder (low value)
   2.   Shrub to 3 m; highly branched; bipinnate, compact leaves; low
      elev tropics

ACACIA XANTHOPHLOEA Benth.
    1.   Africa, India; ornamental, fodder
    2.   Tree to 20 m; handsome, yellow green; spiny; bipinnate;
          frost-free tropics

ACACIA SPP Many of the Acacia spp. (about 800 Australian, 200 African, 200
    American--basically subgeneric and quite distinct) are multipurpose
    shrubs or trees with agroforestry potential, including: A.
    argyrodendron, A. brevispica, A. currasavica, A. cyperophylla, A.
    deanei, A. doratoxylon, A. flavescens, A. gerrardii, A. harpophylla,
    A. kempeana, A. leptocarpa, A. macrothyrsa, A. mellifera, A.
    murrayana, A. nigrescens, A. oswaldii, A. shirleyi, A. sparsiflora,
    A. stenophylla, A. sutherlandii, A. tetragonophylla, A. tephrina, and
    A. tumida

AESCHYNOMENE ELAPHROXYLON (Guill. and Perr.) Taub. "Ambatch", "Pith Tree",
    "Balsawood tree" (MIMOSOIDEAE: LEGUMINOSAE)
    1.   Tropical Africa, now widespread; light corky wood, fodder
          (leaves)
    2.   Shrub or small tree to 9 m with sticky hairs; mesic and swampy
          tropics to 2000 m; = Herminiera elaphroxylon

ALBIZIA CHINENSIS (Osb.) Merr. (MIMOSOIDEAE: LEGUMINOSAE)
    1.   India; In N India to 1300 m; timber, fodder (low DMD 38%, mildly
          toxic), shade tree
    2.   Tree to 15 m; deciduous; rapid growth; mesic and dry subtropics

ALBIZIA LEBBEK (L.) Benth. "Siris", "Woman's Tongue"
    1.   Asia, Africa; worldwide now; ornamental, fuelwood, fodder (DMD
          50%, new growth toxic), furniture wood
    2.   Tree to 25 m; Wide adaptability, dry-humid tropics, to 1500 m
          elev, to 600 mm rainfall; growth to 8 m in 8 yrs.

ALBIZIA ODORATISSIMA Benth.
    1.   Nepal, India; ornamental, fodder (fair quality)
    2.   Tree to 25 m; humid subtropics, to 1500 m

ALBIZIA PROCERA (Roxb.) Benth. "Forest Siris", "Safed Siris"
    1.   India, SE Asia and N Australia; now widespread; lumber,
          fuelwood, furniture, shade tree, tannin, soil rehabilitation
    2.   Tree to 20 m; humid sub-tropics, to 1800 m; not frost tol;
          pollards well

ALBIZIA SPP.: Other multipurpose species of possible use in agroforestry
    include A. adianthifolia, A. amara, A. basaltica, A. harveyi, A.
    lophantha, A. richardiana, A. stipulata and A. toona.

ALNUS ACUMINATA O. Kuntze (BETULACEAE)
    1.   C America; firewood, timber, craftwood
    2.   Tree to 25 m; cool tropical highlands; low tol of heat or
          drought, coppices well

ALNUS GLUTINOSA (L.) Gaertn. "Black alder"
1.   Europe, Asia; now widespread; fuelwood, craftwood, soil
     stabilization
2.   Tree to 40 m; temperate or subtropical; not drought tol

ALNUS NEPALENSIS D. Don. "Nepal alder"
1.   Himalayas; firewood (low density), furniture, timber, fodder
2.   Tree to 30 m; cool tropic highlands; subhumid, fast-growing

ALNUS RUBRA BONG. "Red alder"
1.   N America; fuelwood, furniture (s.g.=0.4), soil improvement,
     nurse tree, fodder
2.   Tree to 25 m; temperate

ALLOCASUARINA LITTORALIS (Salisb.) L. Johnson. "Black sheoak"
     (CASUARINACEAE)
1.   NE to SE Australia; wood for turnery, fuelwood, charcoal
2.   Shrub or tree to 12 m; wide ranging in latitude and altitude;
     humid substropics, but drought tol

BAPHIA NITIDA Lodd. "Camwood" (PAPILIONOIDEAE; LEGUMINOSAE)
1.   W Africa; wood for dyes, fodder (browse), construction,
     medicinal, ornamental, fencerows
2.   Shrub to 3 m or tree to 10 m; humid tropical forests

CAJANUS CAJAN (L.) Millsp. "Pigeon Pea", "Dhal", "Catjang"
     (PAPILIONOIDEAE; LEGUMINOSAE)
1.   Africa (cult < 2000 BC) and India; now worldwide; food,
     medicinal, green manure, fodder (rare, not persistent),
     windbreak, honey
2.   Variably perennial shrub to 4 m; annual types for food; dry
     tropics, low to midland; not tol of grazing, frost or fire

CALLIANDRA CALOTHYRSUS Meissn. "Kaliandra" (MIMOSOIDEAE: LEGUMINOSAE)
1.   C and S America; fuelwood, green manure, poor fodder (DMD 41%),
     ornamental
2.   Shrub to 7 m; moist tropics, cold tol; rapid growth; acid tol,
     poor on alkaline soils

CALLIANDRA ERIOPHYLLA Benth. "False Mesquite"
1.   Mexico to USA; browse, fuelwood
2.   Shrub to 3 m, dense; thornless

CASUARINA CUNNINGHAMIANA Mig. "River she-oak" (CASUARINACEAE)
1.   Australia; widespread; firewood, charcoal, dense timber
2.   Tree to 30 m; tropical to subtropical; subhumid, saline tol

CASUARINA EQUISETIFOLIA L.
1.   Australia; now worldwide; fuelwood (s.g.=0.95), charcoal ("best
     in world"), windbreak, dune stabilization, posts
2.   Tree to 30 m; warm tropics, coastal, typhoon and saline tol

CASUARINA GLAUCA SIEB. ex Spreng.
  1.   NE Australia; firewood, charcoal, fencing, windbreak
  2.   Tree to 20 m; warm temperate, subtropical; coastal, salt tol;
       root suckers, can be weedy

CASUARINA JUNGHUNIANA Miq.
  1.   Indonesia; now SE Asian (hybrid with C. equisetifolia);
       postwood, timber, fuelwood, charcoal, windbreak
  2.   Tree to 25 m; moist tropical, coastal; saline and acid tol

CEANOTHUS SPP. "Buckthorns" (RHAMNACEAE)
  1.   Americas; ornamental, fodder (browse), soil amelioration
  2.   Shrubs to 5 m; drought hardy; diverse habitats; species hybrids
       common among the 50 species in horticulture

CHAMAECYTISUS PALMENSIS (Christ) Bisby et Nicholis. "Tagasaste", "Tree
  Lucerne" (PAPILIONOIDEAE: LEGUMINOSAE)
  1.   Canary Islands, to NZ; hedges, fodder (high DMD 70%), bee
       pasture
  2.   Shrub to 6 m; temperate, frost tol to -10C; drought tol, not tol
       of acid soils

CODARIOCALYX GYRANS (L.) Hassk. "Telegraph plant" (PAPILIONOIDEAE:
  LEGUMINOSAE)
  1.   Indo-Malaysia and Philippines; green manure, browse fodder
  2.   Shrubs and sub-shrubs; tol of poor drainage

DALBERGIA SISSOO Roxb. "Sissoo", "Indian teakwood", "Tali"
  (PAPILIONOIDEAE: LEGUMINOSAE)
  1.   India; timber, fuelwood, shade tree, fodder (poor)
  2.   Tree to 30 m; tropics and subtropics to 1200 m; mesic, to 800 mm
       rain

DESMANTHUS VIRGATUS (L.) Willd. "Donkey Bean" (MIMOSOIDEAE: LEGUMINOSAE)
  1.   S and C America; now widespread worldwide; browse fodder (DMD
       53%)
  2.   Subshrub to 3 m; short-lived perennial; aggressive, coppices and
       reseeds well; unarmed; mesic tropics but drought tol; not acid
       tol

DESMODIUM DISCOLOR Vog. "Horse Marmalade" (S Africa) (PAPILIONOIDEAE:
  LEGUMINOSAE)
  1.   S America; now widely distributed; fodder (highly palatable)
  2.   Subshrub to 3 m; woody when mature; subtropical, frost-hardy;
       several related subshrubby species common tropical fodder

ELAEAGNUS ANGUSTIFOLIA L. "Oleaster", "Russian olive" (ELAEAGNACEAE)
  1.   N America; windbreak, ornamental, erosion control
  2.   Shrub to 8 m; temperate, frost and drought tol

ELAEAGNUS SPP.: E. umbellata (Asia) is used as nurse tree and bird feed,
  E. pungens Thunb. (Asia) is ornamental shrub

ERYTHRINA POEPPIGIANA (Walpers) O. F. Cook (PAPILIONOIDEAE; LEGUMINOSAE)
1. S America to Panama; ornamental, fencerow, windbreak, nurse tree
    (coffee), pulpwood, mulch/green manure
2. Trees to 20 m; thorny; mesic to cool tropics

ERYTHRINA VARIEGATA L. "Tiger's claw", "Indian coral tree"
1. C and S America; fencerows, nurse tree (coffee), windbreaks
2. Tree to 25 m; lowland moist tropics, often in swamps; easily
    cloned; one columnar variety used widely in windbreaks

FLEMINGIA MACROPHYLLA (Willd.) Merrill (PAPILIONOIDEAE; LEGUMINOSAE)
1. SE Asia; dyes, fodder (poor), green manure
2. Shrub to 2 m; = F. congesta Roxb., = F. latifolia Benth;
    mesic-wet tropics, moderately shade and acid tol

GLIRICIDIA SEPIUM (Jacq.) Walp. "Madre de cacao", "Quickstick"
    (PAPILIONOIDEAE;LEGUMINOSAE)
1. C America/Mexico; now worldwide; firewood, timber, shade,
    ornamental, fodder (best acceptance if wilted) of high DMD (55%)
2. Tree to 15 m; easily propagated by cuttings; rapid growth; dry
    to mesic tropics to 1000 m elev

    One of the most important agroforestry legumes; highly flexible
    in management and widely adapted. New varieties promise
    improved yields and better form as trees

HIPPOPHAEA RHAMNOIDES L. "Sea buckthorn" (ELAEAGNACEAE)
1. Eurasia; food (fruit), medicinal
2. Shrub to 9 m; temperate, spiny, with high cold tol

INGA VERA (L.) Britton (PAPILIONOIDEAE; LEGUMINOSAE)
1. C America, Caribbean; fuelwood, timber (dense), shade, honey
2. Tree to 20 m; humid tropics, lowlands; A genus with many other
    species of agroforestry potential, including I. jiniquil

LEUCAENA DIVERSIFOLIA (Schlecht.) Benth. "Red Leucaena" (MIMOSOIDEAE:
    LEGUMINOSAE)
1. C America, Mexico; recently worldwide; coffee shade, fuelwood,
    postwood, fodder
2. Tree to 18 m; mesic tropics, midlands to highlands; fairly acid
    tol; rapid growth; high psyllid tol

LEUCAENA LEUCOCEPHALA (Lam.) de Wit. "Leucaena", "Ipil-ipil", "Lamtoro"
1. C America, Mexico; worldwide; fodder (high DMD, 60-70%),
    fuelwood, shade, pulpwood, postwood, lumber, food
2. Tree to 20 m; dry to mesic tropics; not acid tol, growth slow in
    highlands; widely studied and planted, fast growth

LEUCAENA SPP.
1. N America; uncommon internationally; food, fodder, fuelwood
2. 10 other species of shrubs and trees to 20 m; All can be
    hybridized with L. leucocephala; dry to mesic, lowland to
    highland; several highly resistant to psyllids (e.g., L.
    pallida, L. collinsii, L. esculenta)

Probably the most versatile agroforestry genus; many varieties
and over 50 species hybrids becoming available; wide dispersion
of single varieties discouraged due to self-fertility and
potential pest problems.

<u>MEDICAGO ARBOREA</u> L. "Tree Medic", "Cytisus Shrub" (PAPILIONOIDEAE;
LEGUMINOSAE)
1. Greece; now common in Mediterranean; described 100 AD as
    valuable goat fodder
2. Small shrub to 4 m; greyish, silky hairs; sub-temperate, not
    hardy against severe frost; drought tol

<u>MIMOSA SCABRELLA</u> Benth. (MIMOSOIDEAE; LEGUMINOSAE)
1. SE Brazil and Argentina; fuelwood, pulpwood, nurse tree,
    ornamental
2. Erect tree to 12 m; thornless, rapid growth; cool moist tropics
    and subtropics

<u>OUGEINIA OOJEINENSIS</u> (Roxb.) Hochr. (PAPILIONOIDEAE; LEGUMINOSAE)
1. N India; wood for implements, fodder
2. Tree to 14 m; slow growth (2 m in 6 yrs); to 1200 m in N India,
    frost and drought tol

<u>PARKIA SPP.</u>
Several <u>Parkia</u> <u>spp.</u> are important food crops (pods/seeds) with some
use as shade and fodder (e.g., <u>P.</u> <u>clappertoniana</u>, <u>P.</u> <u>filicoidea</u>, <u>P.</u>
<u>javanica</u>)

<u>PERISERIANTHES FALCATARIA</u> (L.) Nielsen
1. Indonesia, New Guinea; widespread in humid tropics; timber (low
    s.g.=0.33), pulpwood, soil improvement
2. Tree to 40 m; also treated as <u>Albizia</u> <u>falcataria</u>; humid tropics
    to 1000 mm min; midlands

<u>PITHECELLOBIUM DULCE</u> (Roxb.) Benth. "Kamachili", "Dutch Tamarind"
(MIMOSOIDEAE; LEGUMINOSAE)
1. America; now international; fuelwood, shade, ornamental,
    postwood, fodder (pods, seeds)
2. Tree to 20 m; thorny; very wide adaptability from dry to humid
    tropics and sub-temperate regions

<u>PONGAMIA PINNATA</u> (L.) Pierre. "Karang", "Derris" (PAPILIONOIDEAE;
LEGUMINOSAE)
1. India to China, Australia, and Malaysia; oil seeds, shade tree,
    medicinal, firewood, fodder (lopped, dried; fresh leaves
    distasteful; DMD 50%), craftwood, bark for fiber, seed cakes
2. Small tree to 8 m; = <u>Derris</u> <u>indica</u>; mesic tropics (to 600 mm);
    salt tol

<u>PROSOPIS ALBA/CHILENSIS</u> (<u>P. alba</u> Griseb., <u>P.</u> <u>chilensis</u> (Mol.) Stuntz)
(MIMOSOIDEAE; LEGUMINOSAE)
1. S America; firewood, timber, fodder (pods), shade
2. Trees to 15 m; thorny; hot dry tropics, to 100 mm, also to
    highlands (3000 m in Argentina)

PROSOPIS CINERARIA (L.) Druce. "Khejri"
      1.   India; used before 1000 BC; firewood, charcoal, green manure,
            postwood, fodder (but high tannis and low DMD, 40%)
      2.   Tree to 9 m; thorny; hot dry tropics to 10 mm, normally 500-800
            mm; light demanding

PROSOPIS PALLIDA/JULIFLORA "COMPLEX" (P. pallida Humb. & Bon. ex Willd.
      and P. juliflora (Swartz) DC). "Algaroba", "Ironwood", "Keawe"
      (Hawaii)
      1.   C and S America; now widespread; fuelwood, excellent charcoal,
            fodder (pods), honey, wood, coastal soil stabilization
      2.   Trees to 15 m; thorny (segreg.); hot dry tropics (to 200 mm);
            saline tol

PROSOPIS SPP.
      Other P. spp. providing fuelwood and animal feed, normally pods,
      include P. glandulosa Torr. (weedy), P. spicigera L. and P.
      tamarugo F. Phil.

PTEROCARPUS ERINACEUS Poir. "African Rosewood", "Apepe" (MIMOSOIDEAE;
      LEGUMINOSAE)
      1.   W Africa; wood for tools and posts, fodder (foliage), dyes and
            tanning, afforestation
      2.   Tree to 15 m; mesic tropics; good on shallow soils

PTEROCARPUS MARSUPIUM Roxb.
      1.   India; fodder (foliage), fuelwood, timber
      2.   Tree to 30 m; coppices and pollards well; mesic tropics, some
            frost tol

PURSHIA SPP. "Bitter brush", "Antelope brush" (ROSACEAE)
      1.   N America; browse fodder for game, soil erosion control
      2.   Shrubs to 3 m; temperate, high cold tol

ROBINIA PSEUDOACACIA L. "Black Locust" (PAPILIONOIDEAE; LEGUMINOSAE)
      1.   N America; now widespread in highland tropics; fuelwood,
            ornamental, honey, reforestation, land stabilization, nurse
            tree, fodder of low value (27% DMD) and some toxicity
      2.   Tree to 20 m; fast growth; highland tropics (to 3000 m); one of
            the few temperate N-fixing legume trees; forms thickets

SAMANEA SAMAN (Jacq.) Merrill. "Raintree", "Cow Tamarind" (MIMOSOIDEAE;
      LEGUMINOSAE)
      1.   India; now widespread especially in Africa and N India, fodder,
            green manure, possible pulpwood
      2.   Annual shrub to 6 m; thorny; wet or saline tropics; weedy; young
            leaves used as cattle fodder

SESBANIA GRANDIFLORA (L.) Poir. "W Indian Pea Tree", "Katurai" (Phil),
      "Agati" (Indonesia), "Gallito" (Caribbean)
      1.   Indonesia and Australia; now worldwide; food (flowers, pods,
            leaves); fodder (foliage); pulpwood, ornamental, nurse tree
      2.   Tree to 10 m; = S. formosa F. Muell; short-lived; fast growth;
            slow foliage regeneration; mesic tropics (> 1000 mm); tol of
            waterlogging

SESBANIA SESBAN (L.) Merrill. "Sesban"
1.   Tropical Africa (widespread), Asia; green manure, fodder, fiber
2.   Shrub to 5 m; = S. aegyptiaca (Poir) Pers.; fast growth; moist
      tropics; saline and flooding tol

SESBANIA SPP.
      Other S. spp. in use or under evaluation are subshrubs or shrubs,
      incl. S. cannabina and S. speciosa

SHEPHERDIA ARGENTEA Nutt. "Buffalo berry" (ELAEAGNACEAE)
1.   N America; fodder (fruits, foliage), soil erosion control
2.   Shrub to 5 m; temperate, high drought and cold tol

TAMARINDUS INDICA L. "Tamarind" (PAPILIONOIDEAE; LEGUMINOSAE)
1.   Tropical Asia and Africa; food (fruit), shade, medicinal,
      fuelwood and charcoal
2.   Tree to 25 m; tropical, drought tol, slow growing

5. AGRICULTURE, FORESTRY, AND WILDLIFE RESOURCES...PERSPECTIVES FROM THE
WESTERN HEMISPHERE

J. F. Eisenberg* and L. D. Harris

School of Forest Resources and Conservation and *Florida State Museum,
University of Florida, Gainesville, FL 32611

5.1. INTRODUCTION

Approximately ten thousand years before the present, humans in Europe
and Asia began what is known as the Neolithic Revolution. In short, the
domestication of plants and animals accelerated within cultures in the Old
World and the sciences of agriculture and animal husbandry were born.
From simple village life in the Old World, civilizations developed with
city-states setting policy with respect to productivity and land usage.
The old civilizations waxed and waned only to give birth to new ones. By
the 15th century most humans in Europe and Asia tended crops or flocks of
domesticated animals on land that did not belong to them but rather
belonged to some privileged caste (the exception, of course, was city
dwellers not directly engaged in agricultural production).

The discovery of the New World in the late 15th century began a phase
of exploration and accelerated trade. The colonization of North and South
America by Europeans involved Spain, Portugal, Holland, France, and
England and had many adverse impacts on indigenous cultures. Although
Spain and Portugal established early footholds in the north, their
principal interests and settlements were in Central and South America.
The Dutch, English, and French concentrated on North America and
eventually English colonial policies came to dominate this continent prior
to the American Revolution. Many of the new settlers from western Europe
were dissatisfied with land ownership by a few privileged individuals.
Thus, early on in North America and especially in the northeast,
agricultural efforts centered on small landholdings held by individual
families. Although plantation agriculture developed in the southeast
based upon slave labor, the concept of family landholdings and the family
farm soon came to dominate the North American continent (2).

Although most of the world's population now depends on agriculture
and animal husbandry for its sustenance, wild game and even insects are
utilized by rural populations all over the world. For example, over 20
grams of protein per day consumed by individuals of the Isoko farming
community in Nigeria derive from the wilds (Table 1). In a Brazilian
frontier community, nearly 3,000 kg of wild game are taken on an annual
basis (Table 2). Wildlife is still important in rural communities
worldwide, and could be utilized much more extensively on a sustained-
yield basis.

5.1.1. The North American Experience

As immigrants colonized the west of North America there was

Table 1. Average daily consumption of domestic and wild animals and fish and the amount of protein provided in the diet of the Isoko farmers of the Niger delta, Nigeria (21). Total vegetable protein consumed daily was 26.1 g.

Food	Edible Portion (g)	Protein Content (g)
Fish (fresh)	18	3.0
(dried)	18	9.0
Monkey (fresh)	8	1.2
(dried)	7	3.2
Goat	6	1.0
Pangolin and Porcupine	5	0.9
Grasscutter and giant rat	3	0.6
African snail	3	0.6
Palm weevils	1	0.1
Frogs	3	0.6
Total		20.0

Table 2. Game taken near Agrovila Nova Fronteira, Brazil (September 1973 - August 1974) (27).

Common name	Scientific name	No. Killed	Total (Kg)	Total (%)
South American tapir	Tapirus terrestris	8	1,180	36.7
White-lipped peccary	Tayassu pecari	52	1,145	35.6
Brocket deer	Mazama americana	15	415	12.9
Collared peccary	Tayassu tajacu	8	110	3.4
Jaguar	Felis onca	1	80	2.5
Tortoise	Geochelone sp.	21	62	1.9
Paca	Agouti paca	8	59	1.8
Rabbit	Sylvilagus brasiliensis	31	33	1.0
Agouti	Dasyprocta sp.	14	32	1.0
Puma	Felis concolor	1	24	0.7
Nine-banded armadillo	Dasypus novenicinctus	6	16	0.5
Giant anteater	Myrmecophaga tridactyla	1	15	0.5

inevitably much abuse of the land. By the mid-19th century vast herds of native big game had been eliminated from the central plains of Canada and the United States. Timber companies opening up the great northern forests adopted a cut-and-run approach with little thought given to reforestation.

It was only in the early 20th century, under the administration of Theodore Roosevelt, that the first national parks and forests were established in the United States and forestry as a science was transplanted from Europe to North America. At this point it became national policy to encourage reforestation and fire management, and the concept of sustained yield based on a long rotation cutting scheme was introduced (26).

Whereas prior to this time each of the states was responsible for its wild game, the passage of the Lacy Act in 1901 permitted the federal government to regulate trade in wild game products between and among states, for the first time offering some federal regulation to the exploitation of wildlife resources. Schenck established the first forestry school on the Biltmore Estate near Ashville, North Carolina. Biltmore graduates became the first cadre of trained foresters to enter the private and public sectors of forest management. Aldo Leopold published his "Principles of Game Management" in 1933 and founded the first school of wildlife management at the University of Wisconsin (16).

Since the family farm was the cornerstone of American agriculture, and further, since these units were often largely self-sufficient in agricultural produce and firewood, the concommitant management of small farms for wildlife resources also became firmly engrained in the United States and Canada. The fundamental management concept was that, if the natural history of the game animals was known, the numbers that could be harvested on an annual basis without depleting the stock for the next generation could be estimated by appropriate censusing techniques. Farmers could also lease woodlots or fallow lands to hunting clubs or individual hunters and thereby derive additional income. Much effort was expended in aiding individual farm families to plan for this multiple use of the land, and it was highly successful in some of the midwestern states. Parallel developments involving the sustained use of wildlife on public lands were administered by state and federal agencies. Thus, a mixed land use plan evolved, with its roots in Europe and transplanted to the United States, involving economic incentives for individual, small landholders and legal and economic incentives for the management of state- and federal-held lands.

5.1.2. <u>South America</u>

Much of South America lies in the geographically defined tropics, between the Tropics of Cancer and Capricorn. Agricultural methods of the Spaniards and Portuguese were not easily transplanted to the tropical latitudes. The Spanish, through conquest of the highlands of Central and South America, were able to transplant a part of their agricultural practices to these zones, but the true tropics proved intractable. The Portuguese eventually developed plantation agriculture based on slave labor in eastern Brazil. But the colonization of South America based on a modified European scheme of agriculture really was practical only in extreme southeastern Brazil and in Argentina and Chile where the temperate climate was more receptive (2).

In the temperate zones of both the northern and southern hemisphere, seasons are sharply defined. Furthermore, during the summer, day length is greatly extended. The result is that, all things being equal, an annual pulse of productivity occurs allowing a farmer to capitalize on the benign summer climate and maximize agricultural productivity. In the wet tropics, generally poorer quality soils, a constant photoperiod, and often tremendous rainfall with cloud cover do not permit strongly pulsed productivity but, rather, primary productivity is more evenly spread over the year. These large differences in soils, climate, vegetative competition, and a prevalence of plant pathogens and consumers (pests) did not permit the ready transplantation of agriculture from western Europe to the tropics (2).

5.2. A PERSPECTIVE ON THE HABITATS OF THE WESTERN HEMISPHERE

The land in the western hemisphere extends from 80°N latitude to 50°S latitude. Vast mountain ranges, mainly running north and south, offer a variety of climates at all latitudes. This essay will only be concerned with forested habitats. Extreme southern South America and much of North America, especially in the west, are dominated by gymnosperms. The forests of firs and pines offer interesting challenges to wildlife managers. Compared with a broad-leaf deciduous forest of angiosperms, the coniferous forests offer fewer opportunities for the exploitation by higher vertebrates. True, apical buds may be cropped by browsers and cones may be exploited for seeds, but conifers themselves are well-protected against both vertebrate and invertebrate predators. It is the understory, composed mainly of angiospermous shrubs, that is the focus for management of wildlife productivity in conifer forests.

On the other hand, broad-leaved angiosperm forests underwent their evolution at the time that mammals and birds were also rapidly evolving. As a consequence, interesting co-evolutionary symbioses were established, whereby pollination and seed dispersal are often accomplished by vertebrates. Although angiosperm forests dominate the eastern United States and parts of extreme temperate South America, the real stronghold of the angiosperms is in the geographically defined tropics. However, the tropics should not be thought of as a habitat uniformly covered with multistratal tropical evergreen forests. Indeed, rain shadow (orographic) effects of mountains and the course of the prevailing winds on a continental basis, in no small measure determine rainfall, and this in turn determines the form of vegetative cover. Vast tracts of the tropics are covered with dry deciduous forests, eventually giving way to savanna formations, especially where fire has played a role in the history of land use. Their is a tremendous challenge in utilizing both tropical dry and wet forests as renewable resources.

The economic exploitation of temperate zone coniferous and hardwood forests has a long history, and with the rise of forestry as a science, most of the major trees in the northern hemisphere have been well-studied and their life cycles understood. On the other hand, in the tropical forests we have barely begun. In a forest stand in the state of Washington, on the west side of the Cascade mountains, below 700 m elevation, only three tree species dominate--western hemlock (_Tsuga heterophylla_), western red cedar (_Thuja plicata_), and Douglas-fir (_Pseudotsuga menziesii_). But in lowland tropical rain forests, such as the 15 square kilometer tract known as Barro Colorado Island in Panama, over 800 species of woody plants may be identified. We have only begun to understand the life cycle and ecology of a few tropical hardwoods, and we can only guess at most of their potential economic and ecological values (19).

5.3. LANDSCAPE ECOLOGY AND TROPICAL FORESTS

Over the broad span of the geographically defined tropics, climates vary widely in terms of total annual rainfall and its patterning throughout the year. For example, two areas might both receive 1200 mm of rainfall per year, but one may receive 1000 mm during a five-month period with virtual drought during the remainder, whereas the other may receive 100 mm of precipitation per month. Since the tropics are characterized by minimum fluctuations of temperature at low elevations, rainfall patterning tends to synchronize plant productivity (19).

With rainfall levels exceeding 1,500 mm per year and being evenly distributed, multistratal tropical rain forest tends to characterize the landscape. Such a forest has a bewildering complex of woody plants, lianas, and shrubs adapted to different degrees of shade. In contrast to the temperate zone, natural pure stands of trees are virtually nonexistent. The top stratum is somewhat more shade tolerant, and the third still more. Thus, vertical and horizontal heterogeneity in the structure and species present are the dominant features of these forests.

Where there is an evenness of rainfall throughout the year, there may still be peaks of activity, often synchronized by very brief drought periods. However, the annual cycle is characterized by many pulses (28). Also, the great multiplicity of species allows for more even production even though a single species of tree may show a great deal of synchrony, in contrast to the extraordinary pulses so characteristic of higher latitude forests. Where rainfall is more concentrated in time, productivity will also be more concentrated (19).

Considered in a horizontal dimension, landscapes also exhibit a variety of forest types adapted to different degrees of seasonal flooding. Impeded vertical drainage can create areas near rivers that have standing water for appreciable periods of time. These areas may be characterized by a few species, such as palms, that are flood tolerant. Indeed, the flood plains of the great riverine systems of the tropics often show a very reduced woody vegetation but a very high productivity of graminoids, such as in the middle Oronoco River drainage in Venezuela. The meandering pattern and flooding pattern of riverine systems also creates high banks and low banks with vegetation adapted for lower water tables on the high banks and for short periods of flooding on the low banks. These factors interact to create even greater horizontal heterogeneity in structure and form of forests, and also lead to greater temporal asynchrony in production over the landscape as a whole.

Since under favorable conditions multistratal forest will dominate, it is no surprise that much of the productivity occurs in the upper strata. Fruits, flowers and leaves are produced at the higher levels, whereas cover is sparse and productivity is low near the ground. It should also not be surprising then that birds, bats and arboreal mammals characterize closed canopy forests and terrestrial forms are reduced in density and species richness (7).

As mentioned, animals are important in pollinating the angiosperms and in dispersing their seeds, and many intricate systems of interdependency have been studied and shown to be operative in the tropics. Destruction of these animals, therefore, removes irreplacable elements essential for the continued survival of the multiplicity of plant species, and _vice versa_. Emphasis is often placed by wildlife conservation biologists on a small segment of wildlife populations, namely those that are frequently hunted and utilized by indigenous people or exploited for export to more technologically advanced countries for conversion into commodities. These species tend to fall in the middle range of body size, whereas smaller forms are neglected. Yet it is the smaller forms which are more important in pollination and seed dispersion, and toward which conservation efforts should be directed and a greater sensitivity to their role in ecosystems generated.

To illustrate the close coupling of plants and "consumers" in a tropical forest, some of the extensive data from Barro Colorado Island (BCI) in Panama can be used (17, 18). BCI received 2500 mm of rainfall per year, but less than 100 mm on the average falls during January, February and March. There are 57 species per ha of trees greater than 60 cm in circumference; the life expectancy of a tree greater than 60 cm in circumference is 60± 11 years. A gap greater than 20 m^2 is created by tree falls at the rate of one gap per ha per year. The average gap is 65 m^2 in young forest and 90 m^2 in old forest. Fruit production has been estimated at from 2000 to 5000 kg dry weight per ha per year. Fruit fall, measured on the ground, averages only 1000 kg per ha per year; the balance is assumed to be consumption. Leaf fall averages 6500 kg dry weight per ha per year (equivalent to about 6.97 m^2 of leaf area per m^2 of ground). Howler monkeys (<u>Alouatta palliata</u>), sloths (<u>Choloepus</u> and <u>Bradypus</u>), and iguanas (estimated at a population of 10,000 for the island) consume about 22,50 and 50 kg dry weight of leaves per ha per year, respectively. The range of values indicates that vertebrates on BCI consume between 100 and 300 kg dry weight of leaves per ha each year. Insects are estimated to consume another 680 kg dry weight of leaves per ha per year. Finally, it is estimated that insectivorous birds on BCI consume 24 kg dry weight of folivious insects per hectare per year, and that spiders catch insects at the rate of 50 kg dry weight per hectare per year. One could conclude that the predatory efforts of birds and spiders save the forest from complete defoliation.

Of course, leafy plants evolved with their predators, vertebrate or arthropod and, as a result of predation pressures, many plants have developed physical and chemical defenses to protect leaves at critical stages in their development (12, 13, 18, 30). As a generalization, those trees not showing synchronous deciduation employ considerable quantities of secondary compounds to protect their photosynthetic organs. Similarly, seeds are very often rendered toxic by the deposition of secondary compounds. In this case, the plant usually makes the seeds distasteful or quite small while providing a capsule or pericarp with concentrated nutrients, thus attracting the vertebrate or, less commonly the invertebrate frugivore and thereby dispersing viable seeds away from the parent plant (28, 30).

Landscape characteristics are especially critical in our understanding of semideciduous or dry tropical forests, frequently overexploited for charcoal and firewood and more endangered than multistratal tropical evergreen rain forests. These forests are generally less structurally complex than the rain forests. In more xeric areas, where rivers meander through thousands of kilometers of otherwise dry forest, strips of gallery forests occur on high river banks. These strips often support life forms usually found in more mesic habitats (6) and are important corridors for faunal interchange through the otherwise xeric areas. Should gallery forests be interrupted at too many places, the conduit for dispersal of mesic-adapted life forms is broken. Although rather simple in concept, the protection of riparian habitats in either the wet or dry tropics has received very little attention. Indeed, even within multistratal tropical evergreen forests the protection of riparian strips could be key to conserving organismic diversity.

It is often underappreciated that vertebrate species at the top of the trophic ladder, such as raptorial birds, carnivorous mammals and myrmecophageous mammals, exist at very low numerical densities and, by virtue of their trophic strategy, have appreciably large home ranges. Their life history strategies also lead to low fecundity and a highly iteroparous mode of reproduction, all of which bodes for a rather slow recovery from any sort of disturbance. Indeed, the role of top predators in tropical ecosystems is very poorly understood, especially for closed canopy forests, and only a handful of radiotelemetric studies have been conducted (3, 4).

5.4. EXAMPLES OF MIXED LAND USE PATTERNS IN THE TROPICS

The initial European colonization efforts on the islands of the Greater Antilles, beginning in the 15th century, had disastrous consequences for the original ecosystems and the original human inhabitants. Especially crucial was the introduction of European livestock. The rooting activities of pigs and the browsing and grazing pressures of goats and cattle soon made a shambles of the lowland vegetative cover of these islands (2). The culture of the Arawak Indians completely vanished within a century. By the 19th Century, European expansion had begun to take on a more conservative tone. Schenck (26) described the influence of German forestry practices that were adopted in former British colonies. In Burma, for example, attempts were made to superimpose a pattern of teak forestry on an indigenous culture that practiced shifting agriculture ("Taungya", see Lowe, this volume). Teak logs were removed by the government from an area that then became the locus for a controlled resettlement by the original agriculturalists over some five years. The people tended their own crops, including rice, and also replanted young teak trees for eventual harvest by the government some 30 years later. This scheme allowed traditional agricultural practices to coexist with selective logging.

In many areas of the tropics, traditional agricultural practices involved clearing, planting and then abandoning the tract which eventually recovered (8, other chapters in this volume). As an example, Gordon (9) described traditional agricultural practices of the Guaymi Indians of Panama. The Guaymi have an intimate knowledge of the wild plants and game native to their area and not only harvest wild plants for various uses but also capture small game. In addition, they clear forest tracts and plant useful food plants that are then cultivated for several years before the tract is abandoned. These so-called jungle gardens are attractive to small game and, indeed, the resultant mosaic of successional habitats is ideal for the production of agouti (<u>Dasyprocta</u>), paca (<u>Agouti</u>) and brocket deer (<u>Mazama</u>), to name just a few of the more important game mammals. In short, the people take advantage of the attractiveness of their jungle gardens to hunt small game, whose abundance is enhanced by their agricultural practices. The use of small clearings for a limited time, followed by a period of fallow, is, in effect, one of the most useful ways to exploit terrestrial habitats in these latitudes.

Since we have examples before us, it seems reasonable to consider the possibility of a multifaceted approach to the exploitation of tropical forested habitats for human use in contrast to clear-cutting and the establishment of large areas of monocultures of cultivated plants requiring tremendous investments of fossil fuel energy. Ewel (8) provides some creative approaches to tropical agriculture in forested habitats in much greater detail.

5.5. STEPS TOWARDS THE MANAGEMENT OF TERRESTRIAL WILDLIFE RESOURCES IN THE TROPICS

In lowland tropical areas most of the indigenous inhabitants practice some form of horticulture together with hunting and gathering. For various reasons the indigenous horticulturists would would periodically shift their villages and centers of activity. The result was that, as game became depleted in a given area, a shift would not only allow the abandoned area to recover but would also make available new game resources. Large, permanent settlements in the lowland neotropics were only found on sites of productive alluvial soils, renewed annually through riverine deposits, or where the major source of protein derived from fishing or other methods of exploiting riverine or shore environments.

In recent times the opening up of lowland tropical forests to colonists of European origin or colonists utilizing European technology has made hunting more efficient. A headlamp and a shotgun are deadly against nocturnal forms. Furthermore, the tendency of colonists to settle in more or less permanent centers of activity results in the rapid depletion of wild game in the vicinity and no possibility for the game to recover in the absence of legal protection (29). In turn, indigenous cultures are beginning to borrow western technology and incorporate it in their hunting, and, as a consequence of colonization, the ability for indigenous people to move and resettle is more restricted. During this transition phase from a shifting exploitation pattern to permanent settlements, indigenous people may suffer great deprivation as their game supply dwindles (23). Clearly, some compromises must be reached if sustained productive use of lowland tropical habitats is to be achieved.

Extension of the few studies of mammal communities in Venezuela and Panama (3, 5, 7, 29, 30, 32) to other parts of the neotropics indicates wide variation in carrying capacity from habitat to habitat and shifts in increased emphasis on ground mammals as the habitats become more open and savanna-like (3, 5). One of the important game animals of the savanna regions of the neotropics is the capybara (Hydrochaeris), which has been the subject of extensive studies aimed at determining cropping patterns consistent with its utilization as a renewable resource (22). The capybara can coexist on the same range as cattle with little competition. Ready markets for the meat and hides of these semi-aquatic rodents have been established in Venezuela (1).

Much more research is necessary before effective management plans can be developed for the variety of forest areas found within the neotropics. Assumptions based on temperate zone wildlife management experience cannot be directly applied. The density and productivity of species generally varies inversely with body size, yet, as Kleiman et al. (15) point out, some of the important small game species in the neotropics have very low reproductive rates. These generalizations must also be tempered with a knowledge of trophic strategy and phylogenetic affinities (24, 25). In terms of standing crop biomass, the larger species may in fact dominate in spite of their low reproductive potential (14). There now exists the beginnings of a data base upon which to construct management plans for the utilization of terrestrial vertebrates as renewable resources. Model management plans have been developed for the harvesting of the East

African ungulate communities which are pioneer examples of management in tropical savanna habitats (10, 20). Perhaps with continued research we can also make the more complex forest areas as productive for human use.

While scientists and government officials struggle in attempts to develop plans for resource management on a sustained yield basis, the small land holder may wish to consider a mixed land use strategy that not only yields crops but wild game and wood products as well. Such efforts could be augmented by local governments supplying technical advisors, modeling, for example, on the U.S. Department of Agriculture's Wildlife and Fisheries Extension Program. Finally efforts are underway to induct some indigenous game species into semi-captivity and these efforts could be extended to the grass roots level (1, 31). The guinea pig (_Cavia_) was domesticated by the Incas as a food animal. There is no reason why the agouti (_Dasyprocta_) and paca (_Agouti_), could not eventually be brought under domestication in the lowland tropics.

In a pioneering study, Harris (1983) developed a scheme for long-rotation timber harvesting in the forests of northwestern North America which is based on a prescribed spatial arrangement of stands of trees of different ages, allowing climax-adapted species to survive adjacent to areas of earlier successional species, and their inter-connectiveness to facilitate movement of nonaerial animals. While in theory this scheme could be applied to tropical forests, it remains for future research and experimentation to apply and evaluate these principles.

The Man and Biosphere program has revived a time-tested concept in land use planning, involving multiple land use, with attempts to preserve biodiversity. In this case, a central protected habitat is surrounded by several zones of increasing human exploitation farther from the core. For example, from the center outward one would pass from an area of no logging to an area of selective logging to an area of rotational clearcuts. "Agroforestry" systems, as examined in this volume and elsewhere, could provide a key to the effectiveness and acceptance of this approach and its successful implementation at a large scale.

REFERENCES

1. Consejo Nacional de Venezuela. 1976. II Seminario Sobre Chiguires Y Babas. Consejo Nacional de Investigaciones Cientificas y Tecnologicas. Maracay, Venezuela.
2. Crosby, A. W. 1986. Ecological imperialism: the biological expansion of Europe, 900-1900. Cambridge Univ Press, NY. 368 pp.
3. Eisenberg, J. F. 1980. The density and biomass of tropical mammals. In: M. E. Soule and B. A. Wilcox (eds), Conservation biology, an evolutionary-ecological perspective, pp. 35-56. Sinauer Assoc, Inc, Sunderland, MA.
4. Eisenberg, J. F. 1981. The mammalian radiations. Univ Chicago Press, Chicago, IL.
5. Eisenberg, J. F., O'Connell, M. and P. V. August. 1979. Density, productivity, and distribution of mammals in two Venezuelan habitats. In: J. F. Eisenberg (ed), Vertebrate ecology in the northern neotropics, pp. 187-207, Smithsonian Inst Press, Washington, D.C.
6. Eisenberg, J. F. and K. Redford. 1979. A biogeographic analysis of the mammalian fauna of Venezuela. In: J. F. Eisenberg (ed), Vertebrate ecology in the northern neotropics, pp. 31-38. Smithsonian Inst Press, Washington, D.C.

7. Eisenberg, J. F. and R. W. Thorington. 1973. A preliminary analysis of a neotropical mammal fauna. Biotropica 5:150-161.

8. Ewel, J. J. 1986. Designing agricultural ecosystems for the humid tropics. In: R. F. Johnston, P. W. Frank and C. D. Michnew (eds), Ann Rev Ecol Syst, Vol. 17, pp 245-272. Annual Reviews, Inc. Palo Alto, CA.

9. Gordon, B. L. 1982. A Panama forest and shore. Boxwood Press, Pacific Grove, CA.

10. Harris, L. D. 1972. An ecological description of a semi-arid east African ecosystem. Range Sci Dep, Sci Ser 11, Colorado State Univ, Fort Collins, CO.

11. Harris, L. D. 1983. The fragmented forest. Univ Chicago Press, Chicago, IL.

12. Janzen, D. H. 1974. Tropical blackwater rivers, animals, and mast fruiting in the Dipterocarpaceae. Biotropica 6:69-103.

13. Janzen, D. H. 1981. Enterolobium cyclocarpum seed passage rate and survival in horses: Costa Rican Pleistocene seed dispersal agents. Ecology 62:593-601.

14. Kinnaird, M. and J. F. Eisenberg. In press. A consideration of body size, diet, and population biomass for neotropical mammals. In: K. Redford and J. E. Eisenberg (eds), Mammals of the Americas: essays in honor of Ralph M. Wetzel. Univ Cal Press, Berkeley, CA.

15. Kleiman, D. G., Eisenberg, J. F. and E. Meliniak. 1979. Reproductive parameters and productivity of cavimorph rodents. In: J. F. Eisenberg (ed), Vertebrate ecology of the northern neotropics, pp 173-183. Smithsonian Inst Press, Washington, D.C.

16. Leopold, A. 1933. Principles of game management. Scribners, New York, NY.

17. Leigh, E. G., Jr. and N. Smythe. 1978. Leaf production, leaf consumption, and the regulation of folivory on Barro Colorado Island. In: G. G. Montgomery (ed), The ecology of arboreal folivores, pp 33-50. Smithsonian Inst Press, Washington, D.C.

18. Leigh, E. G., Jr. and D. Windsor. 1982. Forest production and regulation of primary consumers on Barro Colorado Island. In: E. G. Leigh, Jr., A. S. Rand and D. M. Windsor (eds), The ecology of a tropical forest, pp 111-122. Smithsonial Inst Press, Washington, D.C.

19. Leigh, E. C., Jr., Rand, A. S. and D. M. Windsor (eds). 1983. The ecology of a tropical forest. Smithsonian Inst Press, Washington, D.C.

20. McNaughton, S. J. and N. J. Geogriadis. 1986. Ecology of African grazing and browsing mammals. In: R. F. Johnston, P. W. Frank and C. D. Michner (eds), Annual Rev Ecol Syst, Vol 17, pp 39-65. Annual Reviews, Inc, Palo Alto, CA.

21. Nicol, B. M. 1953. Protein in the diet of the Isoko tribe in the Niger delta. Proc Nutr Soc 12:66-69.

22. Ojasti, J. 1973. Estudio biologicao del chiguire o capibara. Fondo Nacional de Investigacions Agropecuiarias. Caracas, Venezuela.

23. Redford, K. H. and J. G. Robinson. In press. The game of choice: patterns of Indian and colonist hunting in the neotropics. Amer Anthropol.

24. Robinson, J. G. and K. H. Redford. 1986a. Intrinsic rate of natural increase in neotropical forest mammals: relationship of phylogeny and diet. Oecologia 68:516-520.

25. Robinson, J. G. and K. H. Redford. 1986b. Body size, diet and population density of neotropical forest mammals. Amer Nat 128:665-680.
26. Schenck, C. A. 1974. The birth of forestry in America, Biltmore Forest School 1898-1913. Forest History Society and the Appalachian Consortium, Santa Cruz, CA.
27. Smith, J. H. 1976. Utilization of game along Brazil's transamazon highway. Acta Amazonica 6:455-466.
28. Smythe, N. 1970. Relationships between fruiting seasons and seed dispersal methods in a neotropical forest. Amer Nat 104:25-35.
29. Smythe, N. 1978. The natural history of the Central American agouti (_Dasyprocta punctata_). Smithsonian Contrib Zool 257:1-52.
30. Smythe, N. 1986. Competition and resource partitioning in the guild of neotropical terrestrial frugivorous mammals. In: R. F. Johnston, P. W. Frank and C. D. Michner (eds), Ann Rev Ecol Syst, Vol. 17, pp 169-188. Annual Reviews, Inc, Palo Alto, CA.
31. Smythe, N. In press. The paca (_Cuniculus paca_) as a domestic source of protein for the neotropical, humid lowlands. Appl Animal Behav.
32. Smythe, N., W. E. Glanz and E. G. Leigh, Jr. 1982. Population regulation in some terrestrial frugivores. In: Leigh, E. G., Jr., Rand, A. S. and D. M. Windsor (eds), The ecology of a tropical forest, pp 227-238. Smithsonian Inst Press, Washington, D.C.

6. THE USER PERSPECTIVE AND THE AGROFORESTRY RESEARCH AND ACTION AGENDA

Dianne E. Rocheleau M.

International Council for Research in Agroforestry (ICRAF), P. O. Box 30677, Nairobi, Kenya[1]

6.1. WHY A USER PERSPECTIVE?

Agroforestry has not entered the tropical research and development scene in a vacuum, but rather in a historical context that includes a wealth of recent experience and accumulated knowledge in several related fields. The constellation of relevant concerns includes agricultural and farming systems research and extension, social and community forestry, watershed management and rural development programs in health care, literacy and employment. Each of these fields provides clear lessons which apply directly to the form and content of agroforestry research and development practice in the Third World. Experience in all four areas argues strongly for a user perspective in agroforestry research and action programs. Such an approach requires research and development workers to incorporate the needs, experiences and contributions of rural land users, including farmers, herders, farmworkers, forest dwellers and other rural people.

Both traditional agricultural research and recent farming systems experience have uncovered three key points which influence the success or failure of agricultural programs:

1. Tapping farmer expertise and experience can provide invaluable information and judgement for germplasm selection and breeding criteria, for technology design, for technology evaluation and for site-specific technology adaptation (55, 54, 29). Such information is necessary to guide research station experiments as well as for adaptive research on farm (3).

2. "Successful" technologies can often have a negative effect on specific groups. For example, within households, women might lose income or they might have to work longer hours (67, 66, 45). They could also lose access to cropping system by-products such as crop residues, or edible "weeds". Interhousehold differences within communities can leave the poorest farmers or landless people at a disadvantage when larger farmers adopt a new technology that reduces employment (15). On a regional scale, a change in the cropping system of one farming group may affect seasonal access to fodder for pastoral people that pass through the area (73). Such changes by one group could also leave neighboring farmers in marginal lands without markets or marketing infrastructure for crops abandoned by the first group.

---

[1]Current address: The Ford Foundation, P. O. Box 41081, Silopark House, Nairobi, Kenya

3. "The farmer" is often a farm household or larger unit with multiple objectives, activities, and enterprises on and off farm (36, 20, 1, 59). Motivations of household members, individually and as a group, often include avoidance of risk and juggling of subsistence and cash needs. Allocation of household and individual capital, labor and land resources also reflects these concerns, as well as relative ease of access and control of resources.

Community forestry and social forestry programs also have produced dramatic examples of both failure and success during the last decade. Among the major lessons learned are several applicable to agroforestry:

1. Labor is not the same as participation (33, 24). Many projects based on "popular participation" or "mass mobilization" have in fact been exploitative in nature and have not provided choices, nor have they benefited those who contributed most to establishment of woodlots and other plantations (22, 18). In some cases large groups of villagers have planted seedlings, only to be told later by authorities that they cannot harvest the trees.

2. Communities are not monolithic entities and cannot be treated as such. Even in cases with some degree of local control, community woodlots are often captured by local elites based on wealth or political power (10). Women have provided most of the labor in many projects, while councils of men decide on the species and the management program and reap the major benefits from the trees (37, 35, 33). Individual plantations on private farmland may also be subject to conflicts between different groups at the community level (59, 65).

3. One man's weed may be another woman's livelihood; uselessness is in the eye of the beholder (32). Many agricultural and woodlot schemes have displaced "useless" scrub vegetation that in fact provided other valuable products to particular groups. Women's collection areas for fuel and fodder are often affected, as evidenced in some of the more dramatic encounters of Chipko movement women with "development" projects in fallow and forest lands (44).

4. Subsistence products are not "minor" forest products, and may in fact be major sources of income or sustenance (5, 48, 49). Even when project planners have known of alternative uses of forests and woodlands, they have often underestimated the nutritional, cultural and economic value, and the relative importance of forest products to local people.

5. All trees are not the same, which is to say that species substitution should not be taken lightly nor should it routinely focus on replacement of single commodity values of multipurpose species (73, 2, 64). In many cases a single new species has been chosen and planted in mono-specific stands because of its superior performance for a particular use, or because of its rapid growth and high yields of a specific product. Such species may, however, fail to serve other purposes of major importance to local people (32, 33).

6. "Social forestry" requires viable social contracts (18, 30). Such "contracts" may be between communities and collaborating outside

agencies, among members of participating groups, or between groups of participants and the community at large (32, 33).

Years of research and development programs in watershed management and soil conservation have yielded a number of useful insights about rural resource management in general:

1. Watershed management requires a watershed manager or management unit that is compatible with local practice and organizational structure (26, 16, 12). Planning for the coherent use of a watershed unit cannot be based on the cumulative results of individual decisions taken at farm scale. A manager (or management unit) must be directly accountable to members and to downstream users for the maintenance of whole watershed functions and services. These include water supply, water storage, water regulation and delivery, drainage, maintenance of slope stability and conservation of soil. The latter helps to sustain upland production as well as to prevent sedimentation downstream.

2. Long term conservation strategies should be nested within practices that yield short term (livelihood) benefits for resident participants (12, 74), unless local motivation or subsidies are high enough to offset both local investments and opportunity costs.

3. Patchwork, incomplete and/or inappropriate treatment may be worse than no treatment at all. Well intentioned soil conservation measures, when carried out half-way, or on the wrong site, or incorrectly designed and constructed, can cause gullies, landslips, accelerated sheet erosion, or loss of soil moisture (60, 57). Trees (or AF systems) planted for soil and water conservation purposes can be equally destructive if layout is not properly planned and trees are not planted correctly (72).

4. Widespread adoption of alternative "less effective" treatments may be far more effective than isolated "model farm" adoption of model technologies. For example, 500 hectares of contour planting and vegetation strips on sloping cropland is more desirable than one hectare of perfectly terraced cropland and 499 hectares plowed and planted up and down the slope. The first alternative may be acceptable to farmers, while terracing is not. Technical purism and inflexibility may simply result in a wasted demonstration and extension effort.

5. Relative value attributed (locally) to soil, water, labor and different land uses may determine the range of technology and management options to be considered (60). Labor intensive conservation practices, for example, may not transfer easily to areas where small farmers can earn high wages off farm. On the other hand, in cases where water is scarce, people may be willing to invest more labor and capital in soil conservation structures that also conserve or store water.

Some of the best positive examples for implementation and models of local participation come from action research programs in rural primary health care, adult literacy and alternative local employment opportunities. As reflected in the points listed below, the process, rather than the content, is of special interest for development of participatory research and action methodologies for agroforestry.

1. Local experts, both traditional and "emergent", can often adapt new methods to local conditions better than outsiders. It can be disruptive, as well as costly, to circumvent such people in favor of direct introduction of outside expertise and technology designs. In both Mexico (38) and Kenya local mid-wives and herbalists have played a key role in adapting and disseminating new health care information and practices in rural areas. Stovemaker artesans from the "informal sector" have contributed substantially to the design and "packaging" of improved cookstoves in Kenya (50).

2. Local control of change may matter more than rapid transformation. The process is as important as the initial result if change is to be sustainable. People may need time and a series of small successful changes in order to develop the will and the capability for further change (41).

3. It is usually preferable to build upon and branch from existing technologies, rather than to introduce entirely new technologies. New technologies should be "graftable" onto existing knowledge and practice through a variety of local information and action networks (75).

4. Joint participation between outsiders and rural communities implies dialogue to establish trust and shared goals and to translate and pool knowledge. Facile compromises to meet conflicting or unrelated objectives may simply result in two jobs poorly done, the outcome of a "double lie" (30).

5. Dialogue must take place within communities as well as between the community and outside "catalysts" (25, 41). In some cases, people's responses to outsiders questions and their decisions on community and individual action will change considerably after separate discussion among themselves (69, 59).

6. External constraints that block local initiative may be subject to removal or change, with dramatic results at the local level (11). A change in pricing policy, a marketing co-op, improved access to water, or legal protection of local trees from removal by outsiders, all could open up new opportunities and enable people to take initiatives previously not feasible.

7. Strategies for survival and for positive change can be elicited from "successful" members of disadvantaged groups (11), or from representative members of the client population. Some researchers have engaged rural people in group "think-tank" meetings (36) to improve design and delivery of health care products, while others have used similar approaches to inform technology and project design in agriculture and agroforestry (63, 58).

6.2. WHY A USER PERSPECTIVE PARTICULARLY IN AGROFORESTRY?

In addition to the overall trends and indications for user involvement in rural development and technology research, the breadth of agroforestry also implies a need for a special approach. Assuming a mixture of farming systems and social forestry approaches as a point of departure, an agroforestry user perspective must address the following additional points:

1. the broad scope of the topic (whole-system);
2. the variable scale of the land units involved;
3. the variety of clients and land managers;
4. the ambiguity of land and tree tenure among a multiplicity of clients;
5. the diversity of activities involved (including new categories of work);
6. the combination of production and sustainability (environmental) objectives;
7. the time factor involved in testing and growing trees; and
8. the relative ignorance of researchers as to past and current use of woody plants by farmers, herders and other rural people.

All of these factors place heavy demands on the capability of research and planning teams to collaborate on an equal basis with their rural clients.

6.2.1. <u>Agroforestry for What Uses?</u>

Agroforestry is a land use system, not a commodity. The range of potential practices may apply to cash crops, subsistence crops, animal production and gathered products, as well as to farm infrastructure and to the soil, water and natural vegetation on the site. In contrast to single commodity production approaches, agroforestry has the potential to meet a wide range of rural people's priorities for fulfillment of basic needs (51). Among the major needs which can be met (or at least affected) are: food, water, fuel, cash income, shelter and infrastructure, savings/investment and resources to meet social obligations (Table 1). These needs constitute part of the context of evaluating the full range of "costs" and "benefits" of agroforestry technologies.

Land users evaluate agroforestry and other technologies according to a complex set of criteria that goes far beyond simple economic cost and benefit. They balance considerations of need, preference and multiple use against available resources, required inputs, risk and expected yields. The distribution and timing of all of the above might also make or break a particular technology in the eyes of rural land users.

Rural people's concerns to maintain current production levels into the foreseeable future are an extension of their interest in fulfillment of basic needs. However, the long term maintenance of the natural resource base (soil, water, vegetation) for future production requires more specific attention to history of land use and condition, and to potential improvements in soil conservation, watershed management and management of range and forest lands (56, 60). Any suggestion for improvement implies some vision of alternative future uses of land and other resources. Such concepts of "futures possible" are best presented and discussed in tangible, concrete terms such as landscape and people's livelihoods.

The longer term issues may concern the researchers and national planners more than the local client group. Solutions to these problems will need to be linked to the fulfillment of basic needs (Table 1) over the short and medium term. The extent to which the client group does already recognize and address such problems is critical information for

Table 1. Multiple uses of agroforestry technologies: types of costs and
          benefits.

____

      Cash income (through production and service roles)

            Employment
            Sale of products
            Reduced spending on formerly purchased items
            Direct exchange of products for other goods

      Food supply

            Increased supply
                  total annual supply
                  seasonal distribution
            Improved quality
                  nutrition
                  preference
                  ease of processing/preparation

      Energy supply

            For cooking/home heat/amenity uses
            Increased supply
            Improved quality
            Ease of access/gathering time
            Cost savings (less purchased)

      Shelter/Infrastructure

            Building materials
                  reduced cost
                  available for own use
            Shade
            Protection from wind
            Protection from animals (fencing)
            Definition of boundaries (fencing or markers)

      Savings and investments (S/I)

            Existing form S/I more secure
            Existing form more profitable
            New form S/I

      Raw materials for crafts/cottage industry-for home use or for sale

            Improved supply of currently used material
            Secure future supply of currently used material
            New material supply for same activity
            New material supply for new activity
            New use of existing stock of material

Table 1. (Continued)

    Social production

        Improve existing source of funds, goods, services used to meet
            social obligations
        Intoduces new source of support for social obligations

    Labour profile

        Timing/distribution improved
        Amount reduced
        Returns to labour increased

    Land utilization

        Total output increased
        Type of output improved
                re: priorities of user(s)
                re: range of needs addressed
        Distribution of products/benefits
        Sustainability of use
                Soil (water, fertility, erosion by wind/water)
                Vegetation (amount of cover, quality, regeneration)

    Water balance

        Net yield to dam or water source (stored or flowing)
        Amount stored in soil profile (calendar)
        Amount available for plant growth

---

project design, choice of practices, and nesting of those practices in the
larger landscape.

6.2.2. Agroforestry for Whom (What Users)?

Much of the recent on-farm agricultural and forestry research has
been directed at "target groups" of farm owners and managers. The reality
of rural life is a bit more complex. Any problem that purports to serve
the majority of rural people is by definition dealing with a diverse array
of land users* many of whom are neither owners nor managers of farms.
Even projects specifically geared to target groups such as farmers
will find that their activities and livelihoods may include non-farming
land user roles. Producers depend on a number of items not produced by
them at all, such as gathered products, which may have important
implications for future land use, technology design and labor allocation
on their farms. Farmers' livelihoods may also be inextricably tied to
those of people who are gatherers, processors, merchants, artesans,
farmworkers, herders or forest dwellers. For example, in many Sahelian
agricultural communities, farmers rely heavily on complementary

---

*Land in the broad sense including water, vegetation, wild animals and
other resources on site.

relationships with pastoral nomads whose animals both graze and fertilize croplands in the off season.

Within the group loosely defined as farmers there may also be several types of actors who neither own nor manage the farm as such. Paid farm labor, unpaid (household) farm labor, child care, home management, domestic and commercial processing, gathering of goods for farm household use or sale, and management of herds, small stock or specific plots or plants all constitute essential roles in many farming systems. These are often performed by household members other than the farm owners or managers, or by hired or exchange labor.

Coffee farmers throughout the world provide one major example of both large and smallholders dependent upon farmworkers to harvest their crops. Smallholders, in turn, often depend upon storage and processing facilities owned by largeholders, cooperatives or the state (57, 66). In farming communities where men are the cultivators, women usually process the crops into products for sale or home use.

Not only is there a multiplicity of land user groups but the users may be divided into distinct groups according to a number of criteria. Three criteria which apply to most rural situations are by activity, by tenure (ownership/terms of access) and by unit of organization (Table 2). Each of these criteria defines groups which include horizontal subdivisions as well as vertical hierarchies and which in turn overlap with each other.

Moreover, there is often a separation between spheres of activity and spheres of control between men and women, between age groups and between classes of households. "Management" of specific plants or places is often subdivided between these same groups into major labor input, responsibility to provide goods and services from a given place or plant, and control of the basic resource (means of production). These hierarchies of clients require us to include processing, trade and consumption as well as production processes in the system.

Serving such a complex array of users as clients of agroforestry research and action programs also provides us with an opportunity for adequate and honest treatment of a number of class and gender issues. Distribution of benefits between rich and poor, or men and women are usually ignored by "technical assistance" projects, in spite of the fact that new technologies may change the balance of wealth and control between and within social groups. For example, landless people and women in general tend to depend more heavily on gathering than the population-at-large in farming communities. Whereas wealthier women may gather more on their own land or have easy access to other lands, both landless people and women smallholders often share problems of insecure access to collection areas. A change in the cropping system, a new chemical herbicide, or a change in tree species in bush-fallows may have important "side-effects" on gatherers.

Often agricultural research and extension programs focus on a single "target" group with the hope (even belief) that the result will be a positive impact on regional development, with "trickle-down" to the entire population. If instead we treat all sectors of this whole population as valid clients then we are obliged to deal with the multiplicity and diversity of user groups as well as our accountability to each separately as clients. Whereas "target" and "impact" bring to mind pre-packaged technology "bombs" dropped from above, services rendered to clients implies a range of responses tailored to needs/demands expressed by users in distinct circumstances.

Table 2. Subdivision of land user groups

Land Users...by activity

    A. Producers
            Gatherers
            Hunters
            Herders
            Farmers
                    Large
                    Small
            Farmworkers
    B. Producers
    C. Market vendors
    D. Consumers

Land Users....by rights of access and ownership (applies to trees and/or
    land)

    A. Owner (State, Group, Individual; _de jure_ or _de facto_)
    B. Tenant (Rent paid)
    C. User by permission or exchange agreement
            Continuous
            Regular
            Occasional
    D. Squatters, "Poachers" (illegal users, occupants)

Land Users....by management unit/unit of analysis

    A. Individuals or household sub-groups
            Men, women, children; age group members
    B. Households
            Managed by men, women; small/large;
            young/old; rich/poor
    C. Communities and community groups
            Families, clans, self-help groups
    D. Companies or Cooperatives
    E. Administrative Units
            States, districts, villages, etc.

Examples abound of the need for agroforesters to deal with multiple
users as clients even with respect to single tree species. The case of
Pananao in the Central Mountains of the Dominican Republic illustrates
this point (Fig. 1) in the multiple (and sometimes conflicting) uses of
individual palm trees by men and women. The same tree (or parts thereof)
can be used for fiber (women), cheap construction wood (men) and animal
(hog) feed (men and women).
    In this example the distinct division of control, responsibility (as
providers) and labor extends to spaces and activities as well as to plants
or specific products. Women's processing activities in Pananao require
products from men's fields, herds and woodlands. While women control the
processing enterprise, they do not manage source areas of raw materials.

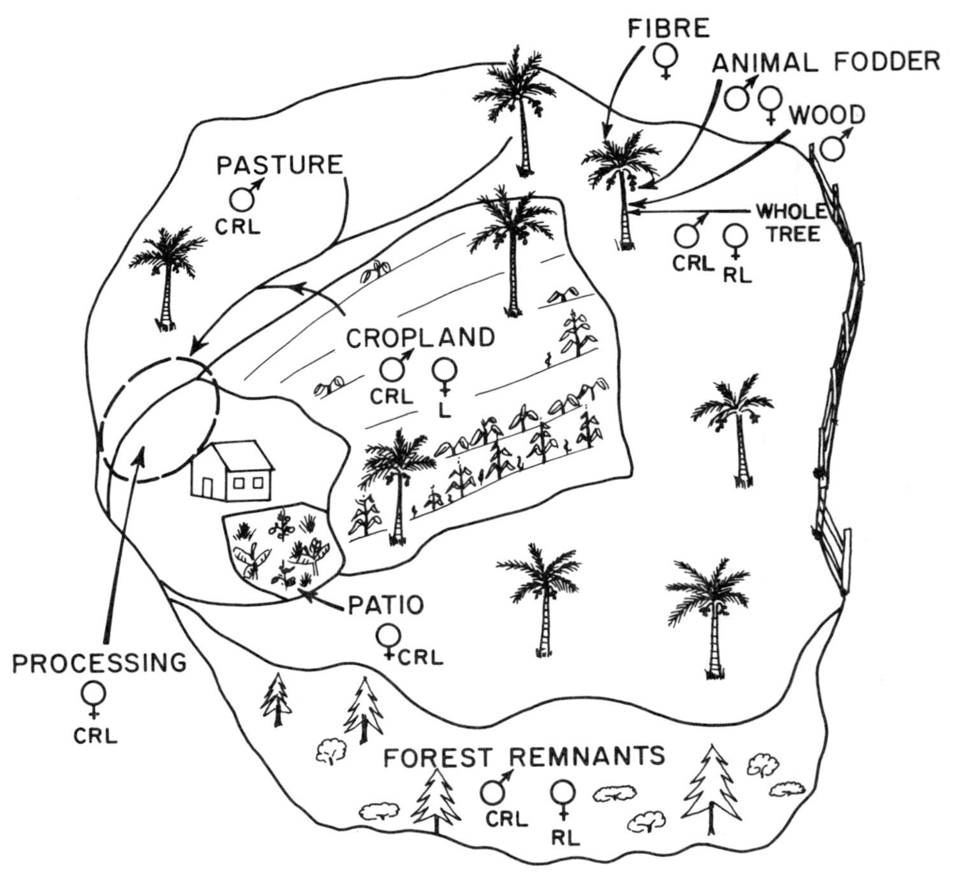

Figure 1. Multipurpose use of land and trees in Pananao (La Sierra, Dominican Republic). It is assumed that both men and women are present in the household. R = responsibility to provide a product thereof to household, L = labor input for establishment, maintenance or harvest, and C = control of resource or process.

Some cassava bread enterprises in the same community have been severely curtailed by fuelwood shortages resulting from rapid conversion of woodlands to cropland and pasture (57). Some women's handicraft enterprises in the vicinity also suffered from raw material shortages when swine fever reduced demand for palm fruit (as hog feed) and men felled the palms for cheap building material or cash.

Agroforestry design in such situations clearly requires consultations with separate client groups to design practices that address each group, whether separately or jointly. Technologies for multiple users can accommodate separate, fully shared, or interlocking (partially shared) use depending on the compatibility of both the uses and the users (Table 3).

Table 3. Agroforestry technology options for multiple users

Parallel technologies (in separate spaces)

    If uses mutally exclusive
        (example: if harvesting fuelwood precludes fruit
        production)
    If users not compatible (on equitable basis)
        (example: if owner denies access to timber trees for
        fuelwood pruning by gatherers)

Interlocking technologies

    If uses or users conditionally compatible
        (examples: Same space or plants, different timing of
        access; Same space, different plants; Same plants,
        different products)

Fully shared, joint technologies

    Compatible (Mutual effect neutral)
        (example: two users harvest same product from boundary
        fence; plenty surplus, fence secure, no competition/
        conflict)
    Compatible and mutually beneficial
        (example: complementary types of labor, share harvest of
        same product)

6.2.3. <u>Agroforestry, Land Users and the Rural Landscape</u>

The rural landscape and the process of "landscape domestication" present a challenging arena for agroforestry design and practice. While this aspect of technology and rural development has been left in the gap between natural resource management, farming systems research and rural organization, it is precisely at this level that many rural people integrate trees, crops and livestock with personal and community needs and objectives. It is also at this scale that many of the most crucial rural land use conflicts occur.

Landscape embodies rural people's ideas and actions over time, in space, in relation to each other and the natural environment. It is a kind of signature in spatial terms that integrates the influence of the past, and provides the geometric point of departure for planning of future land use systems, including agroforestry.

Landscape is the drawing board for integrated agroforestry diagnosis and design beyond the single farm or the individual plot (60, 17). Three aspects of landscape are especially relevant: land units of variable scale, a fairly long time frame for development and change, and complexity of tenure (relative to farms or individual plots).

A relatively long time is required to establish agroforestry systems, let alone to reap the benefits and evaluate their performance. The time constraint on research and development implies a need for special caution mixed with practical imagination of alternative futures. Technology and landscape design must depend heavily on projection of such plans into distinct versions of future development and images of everyday life in the region.

Land and tree tenure are particularly important for tree planters and managers (23), relative to annual cropping which is more ephemeral or animal husbandry which is a more mobile enterprise. Where agroforestry designs apply to several categories of land, land use and plants in a complex landscape, then tenure assumes even greater importance. In addition to careful conceptual screening at the design stage, design also requires a particularly sensitive and responsive research approach that rules out some options early on, and that constantly modifies and improves on the most promising.

Community development cycles (settlement, expansion, diversification/ specialization, land use intensification) will determine in large part the future availability of landscape niches for agroforestry at the community level. Oral history and discussions of "futures possible" may provide some insights into current trends. The choice of appropriate practices and landscape designs requires user involvement from the beginning, even more so in whole community applications.

Within the context of landscape planning and design, a diverse array of agroforestry technologies can address a wide range of land use and production units (Table 1 in Gholz, this volume; see also 60, 61, 52). These units may range from small plots, to farms, watersheds, communal holdings and public lands. The managers may be individuals, households, communal groups of varying size and type, whole ethnic groups, cooperatives, communities, or larger political units. Land use planning at multiple scales requires an integrated social and ecological approach that deals with the division of labor, responsibility, expertise, control, and interests, at intrahousehold, inter-household and community level. To deal with this complexity, development and research workers need to stratify clients by class, sex and household composition, as it affects access to resources and spatial patterns of activity and resource use.

An example from Bhaintan watershed in the Lower Himalaya, Uttar Pradesh State, India (53) illustrates the interplay between multiple users and landscape units in analysis of agroforestry potentials. The landscape sketch (Fig. 2) shows the distinct division in land use and cover which is closely related to tenure. There is a pronounced division of use, control and access to specific landscape features based on sex. The relative share of production (and land use pressure) from a given area also varies by user group (Fig. 3). In turn, the relative importance of particular areas to each user group also varies (Fig. 4). In this case the forest reserve is most heavily used by men, yet it is most important to poor women in terms of its relative contribution to their livelihood. While women's harvest from the forest may be "minor" compared to men's timber off-take, the forest products are major components of their total income. Moreover, poor women's interest in renewable use and sustained yield may be more compatible with national and village level objectives for the commons and forest reserves. Agroforestry technologies for the commons and the forest in this case must deal with the varying interests, intensity of use and needs of each group. Both uses and users must be understood in order to make sense of present landscape and to design systems into the evolving landscape.

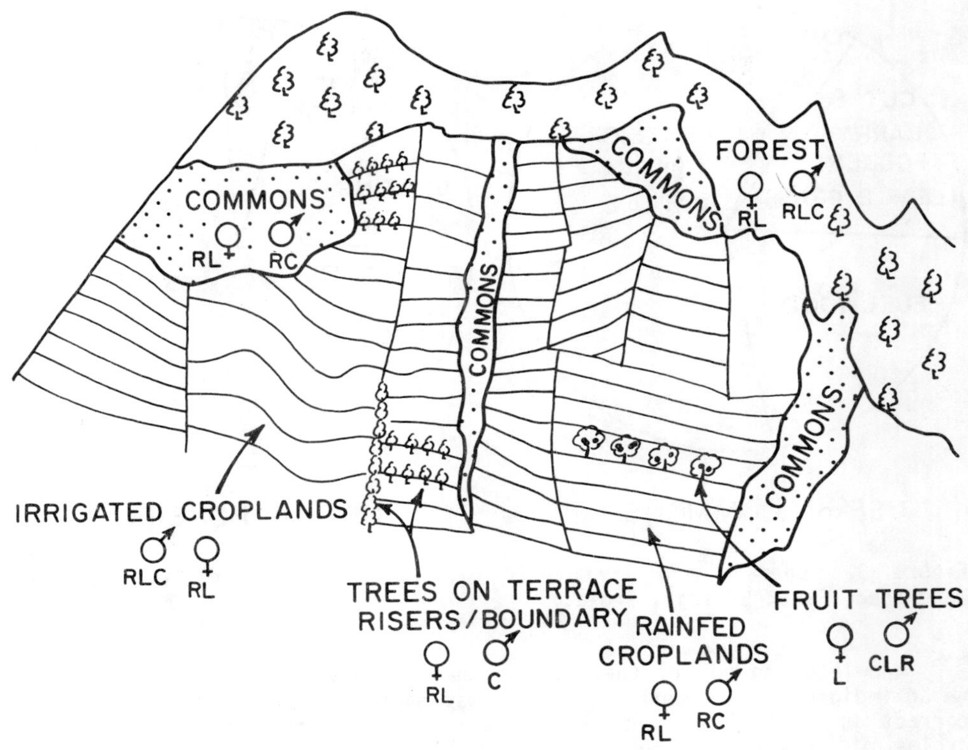

Figure 2. Multiple use of landscape niches and trees in Fakot Village, India (R, L and C as in Figure 1).

6.2.4. Indigenous Knowledge and Agroforestry

Agroforestry as a "formal" science is in a unique position to learn from and to improve upon traditional knowledge and practice and to combine forces with indigenous experimental initiatives and potentials (61). The relative ignorance of the research community about woody plants used by rural people adds a special need for ethnobotanical research to identify promising species (woody and herbaceous) for agroforestry systems, and to understand what is already known about their interaction with soil, animals, other crops and their uses, ownership and management.

One the one hand, there is a tremendous depth of indigenous knowledge about particular traditional systems under very site-specific circumstances (43, 47, 42, 46, 19, 22, 48, 49, 73, 14, 4). Existing knowledge in such systems can span the full range of design and management considerations: site selection, preparation and management; plant selection and/or breeding; plant propagation, establishment and management; plant combinations and spatial arrangements; plant-soil-water interactions; pest management; technologies for processing and use of products; and market conditions at local and regional level.

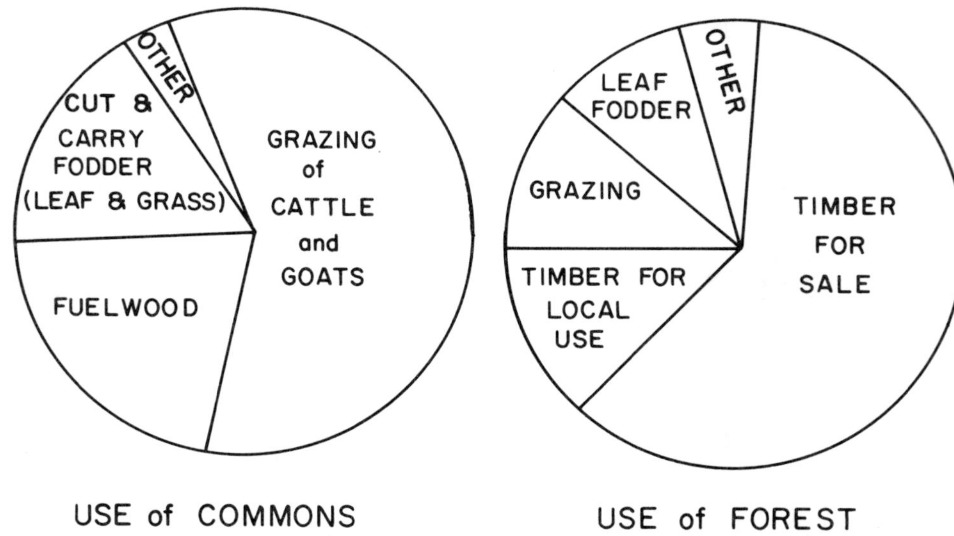

Figure 3. Landuse and livlihoods (by volume harvested) in Bhaintan Watershed, Fakot Village, India.

Knowledge on any of these points may already be coded and evaluated in an indigenous framework (4), or it may simply be embodied in models of correct practice based on the cumulative successful results of prior cycles of experimentation under specific conditions. Such a situation provides two types of information for outside researchers: 1) the coded or otherwise explicit system of indigenous science which explains the interaction of plants, environment and management in the land use system(s), and 2) the system itself, which can be sampled, monitored and manipulated in situ and in model simulations. The translation of information and criteria for judgment and verification between knowledge systems ("ours" and "theirs", or between different bodies of indigenous science) could enrich the ability of both groups to conceive and conduct further experiments (jointly and separately), to design new or improved systems, or to transfer and adapt known systems to new environments (61, 4).

Notwithstanding the richness of knowledge and experience in some existing systems, the overall situation in agroforestry is analogous to that of agriculture during the early Neolithic, when human populations first domesticated and cultivated selected plants (61). Existing and potential agroforestry systems include a particularly diverse array of species, both woody and herbaceous, many of which are wild or only semi-domesticated thus far. In cases where agroforestry is not well developed as such, the local people may still have a wealth of knowledge about useful plant species, including source areas of superior parent material, the ecology of the plant habitat, compatibility with other plants, interaction with animals and insects, growth rate, method of regeneration and response to variation in site conditions and management practice.

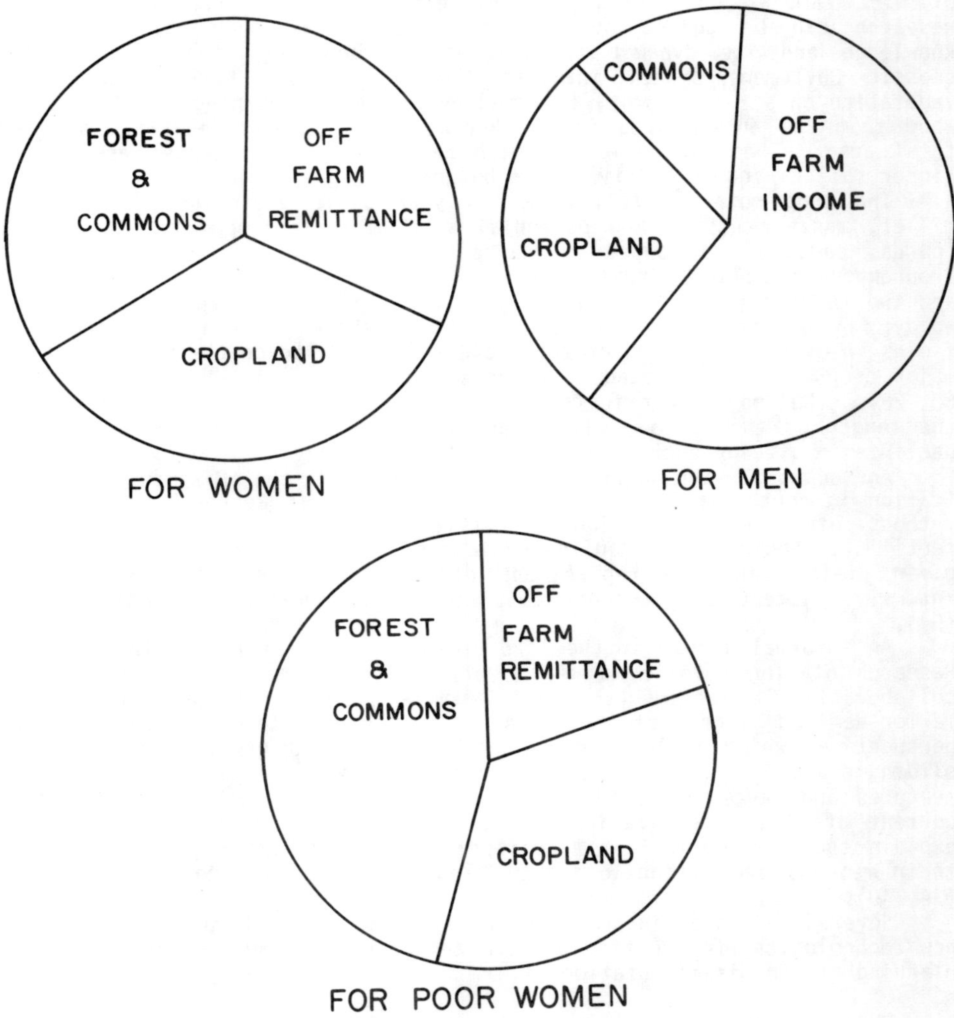

Figure 4. Sources of livlihood (cash and kind) in Fakot Village, India (by relative importance to land user).

Rural people can also play a key role as consumers in deciding the criteria for selection and improvement of germplasm and in judging the likelihood of domesticating particular species or associations within agroforestry systems (61). The dynamic aspects of indigenous knowledge, including capability for innovation and evaluation, can be applied in agroforestry research and design exercises for fitting trees into existing systems, however new, disrupted or traditional they may be.

73

The Chitemene system of shifting cultivation in the "miombo" woodland
of N.E. Zambia constitutes an excellent example of how agroforestry
research can be guided by both the traditional body of indigenous
knowledge and its dynamic application to technology innovation. The
classic Chitemene system involves the felling and harvest of woody
vegetation on a 1-5 ha plot, followed by piling and burning the collected
wood from the entire area on a sub-plot approximately one-fifth of the
total area. The combination of high heat and woody biomass results in
higher soil pH and fertility on the burned plot (39).

The crop rotation follows a six year cycle beginning with finger
millet, maize, cassava and perennial sorghum (intercropped), with yams,
gourds, pumpkin and cowpea on the periphery and/or on termite mounds.
Groundnuts are planted next, followed by cassava maturation and harvest
and two to four years in beans (Fig. 5), after which the plot is left in
woody fallow for several years. Most households maintain at least 4
fields in different stages of the cycle so as to produce the full range of
major crops (millet, ground nut, cassava and beans) in any one year (62,
68, 28). The long term effects of this system on soil fertility vary with
the length of the fallow, with a trend toward shorter fallows and sharp
declines in site productivity (39).

Agricultural and agroforestry research at the Misamfu Research
Station in northeast Zambia has focused on two main approaches: improved
methods of site preparation and fallow for Chitemene (to maintain
fertility, lengthen the cropping cycle and reduce land requirements for a
given yield), and the replacement of the Chitemene system with more
intensive systems of permanent cultivation, with maize as the main crop
(34).

An informal survey of the land users in the vicinity of the Misamfu
Research Station revealed a wealth of information and opportunities for
collaborative experiments on farmer-initiated innovations and
farmer-defined lines of research. The survey incorporated a user
perspective, which included consultation with a variety of users, and a
sliding scale of analysis from region to plot (with emphasis on landscape
features and land use at the farm and community level). The method and
content of these consultations encouraged the clients to draw upon and
explain specific items from traditional bodies of knowledge, as well as
their methods and rationale for developing or adapting new technologies
(34, 40).

Several points of information proved to be critical for the design of
new technologies (for testing on farm and/or on station) and for research
planning at the Misamfu station:

1. While the high input "maize alternative" has been adopted mainly by
    retired civil servants and other relatively wealthy residents of the
    area, many other farmers are actively engaged in experimentation with
    mounding as a way of incorporating plant biomass (usually grass, with
    some tree and shrub parts) into the soil. The mounding of loose
    topsoil over plant biomass has been adapted form the grass mounding
    technology of a neighboring savanna group. It is being tried in
    permanent (or long term) plots planted to beans, cassava and/or a mix
    of fruits, vegetables and specialty crops and it is used in the
    latter part of the cycle on Chitemene plots to prolong the useful
    life of the plot for bean and/or cassava production. Mounds on

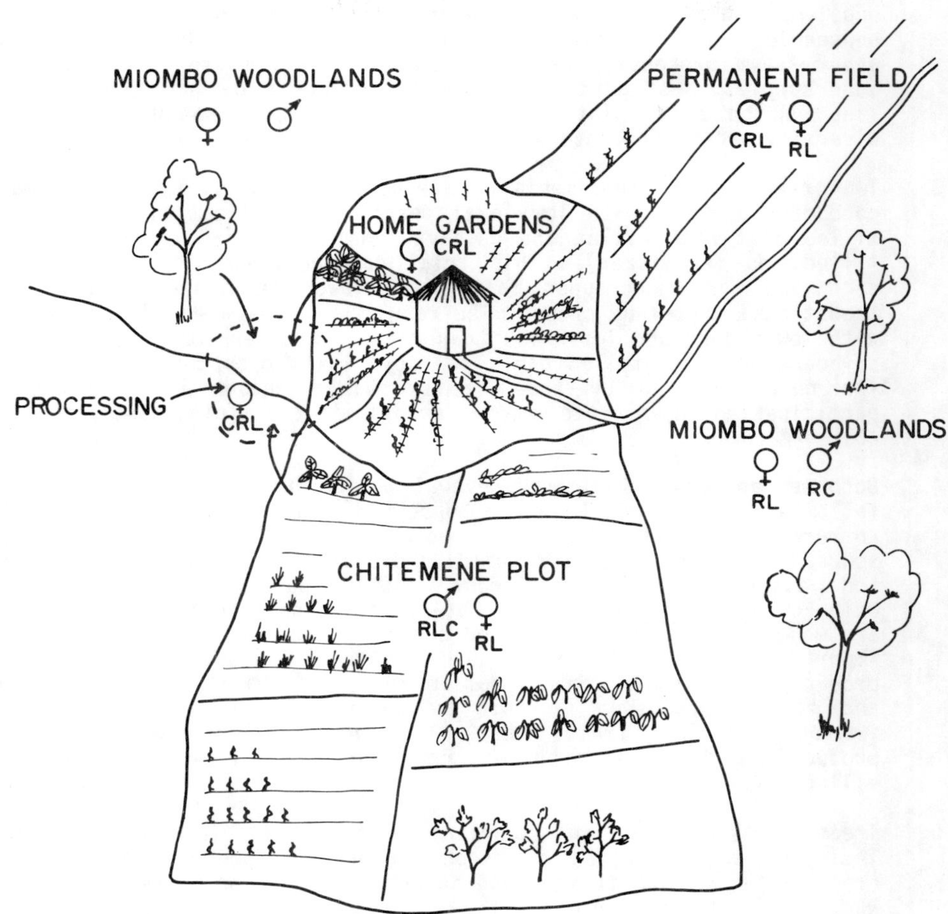

Figure 5. The Chitemene system in northeast Zambia, including new practices observed near Misamfu. Note that women control the millet crop (one of several in the intercrop rotation).

slopes (gentle) usually go across the slope and some are on the contour, which is a conscious soil conservation practice (34, 40, 28).

2. Women's home gardens are becoming increasingly important for food production and cash income and are being diversified to include fruit trees. Some women are experimenting for the first time with tree planting in such gardens. Mounding, raised beds and clean tilled plots are all being tried, with a tendency toward mounding in the larger gardens. Women heads of household rely heavily (even

exclusively) on cassava home gardens for food production to supplement what they can buy with wages. For women without male household labor for Chitemene clearing, the home garden, beer making and cassava processing are important alternatives to earn cash for food supply. Most garden experiments reflect a desire to intensify land use on small plots (limited by male labor, not land) and to diversify processing enterprises (62, 34).

3. Two other farmer experiments were especially noteworthy. One woman conducted a trial with low level fertilizer application on a clean tilled plot with millet and cassava, with a partial control (clean tilled, no fertilizer). This trial combined the site preparation technique for monocropped maize with lower fertilizer levels and traditional Chitemene crops. The result was increased millet yield, with lower cost and less risk than maize. Another woman had planted soybeans on clean-tilled plots for soya milk, prompted by a concern for nutrition and by free seed provided though her daughter's participation in an urban women's program in the mining district (34, 40).

4. Both men and women indicated several important roles of woodland and fallowland products in the household diet and economy (both commercial and subsistence). Woodland and fallow areas are major sources of leafy vegetables (wild) and exclusive sources of mushrooms and caterpillars, which occur mostly on one tree species, <u>Julbernardia</u> <u>paniculata</u> (31). Caterpillars and wild leafy vegetables are major sources of protein and both mushrooms and caterpillars are important sources of cash income for most households. All three products fall within women's domain of responsibility (as providers) and may be processed or sold by women. Timber (men), fuelwood (women) and wild fruits were also cited as important woodland products, with supply problems occurring mainly near towns and old villages (34, 40).

5. Trees as such play an important part in the land use system, including those planted or "kept" in home compounds, fallows and cropland as well as those found in woodland. A considerable body of knowledge and experience exists with respect to both indigenous and exotic, wild and domesticated species. Some farmers had extensive knowledge of exotic fruit tree horticulture, including layering and grafting techniques. Both men and women readily identified their (respective) favorite non-domesticated tree species (by use), those species in short supply, and those that they would consider planting now, or in the future in the event of limited supplies or access (34). Many farmers also provided information on site requirements, potential for management (tolerance to coppicing, pollarding), growth rates (relative) and leafy biomass production (relative) for several species that occur in "miombo" woodland succession (40, 34).

While farmers had not fully incorporated trees into the mounding system, they were willing, even eager, to experiment with "fertilizer trees" to produce high nutrient leafy biomass for mounding. They readily discussed how to nest such trees into mounded plots as well as into fallows and enriched woodland in the surrounding landscape. There is not enough grass for continuous mounding in many areas, particularly near well-situated, crowded villages.

Farmers were often conscious of relative land limits based on proximity to markets, rivers and roads, and most were concerned to define and secure their land rights prior to the immanent return of the mining population to their home area. People's decisions to intensify cropping in place _versus_ to expand their cropland varied mainly with village development cycles and the quality of the village site and services. In many cases people are unwilling to move out and away into outlying woodlands and they intensify production within the context of site-specific land shortages.

The importance of client knowledge, experience and experimental initiatives is best illustrated by a comparative list of the research lines that emerged from this process versus the main research topics previously considered (Table 4).

In many such cases researchers are simply learning what rural people have already discovered. That in itself is critical information for the agroforestry research and policy communities (61, 4). If agroforestry programs are able to absorb such information it may change the whole direction of regional research and development programs including staffing choices, as well as the range of land uses, technologies and species of plants to consider.

If national programs are prepared to "follow the lead" of rural land users, this knowledge of indigenous science and users' initiatives may also alter national agricultural and rural development policy. On the other side, useful information and techniques can best flow from the scientific community to the rural land users once we know what they already know, and what else might be most useful to add to their store of knowledge and tools.

6.2.5. Four Foci of a User Perspective in Agroforestry

The review of the situation facing users and other research and development agents indicates some specific requirements for a user perspective for agroforestry research and action. Such an approach should consider multiple uses and multiple users, landscape as a major focus in a larger context of sliding scale analysis and design, and indigenous knowledge as science. While these are proposed as necessary conditions for understanding and serving users' interests, they are not sufficient to complete the task. The treatment of land users as clients is a critical ingredient in the successful integration of users' concerns into each of the other aspects of analysis, design and action. "Clients" may be seen as passive recipients of services or active participants. The examples presented argue strongly for the latter model to guarantee the best short-term results as well as to build local capabilities for continued AF analysis, design and management.

6.3. THE AGROFORESTRY RESEARCH AND ACTION AGENDA

If the previous points are taken seriously, research and development workers are faced with a tall order. The challenge is to improve/develop technologies based on species and management systems relatively unknown to "formal" science, and to produce results widely applicable across the tropical world, that are adoptable by rural land users in a multitude of diverse environments and circumstances. Furthermore, these multitudes of

Table 4. Introduction of local knowledge into a research plan for
          agroforestry (AF) in northeast Zambia (excerpted from Huxley and
          Rocheleau (34).

____

A.   PREVIOUS RESEARCH DIRECTIONS

     1.   Improvements to Chitemene

          1.1 Modified burning
          1.2. Improved fallow

     2.   AF Alternatives to Chitemene

          - alley cropping of fertilized, limed, maize with three exotic
          tree species: <u>Leucaena leucocephala</u>, <u>Gliricidia sepium</u>, and
          <u>Calliandra callothyrsus</u>.

B.   RESEARCH SUGGESTED BY DIAGNOSIS AND DESIGN SURVEY

     1.   Modification of Chitemene

          1.1 changes in spacing/geometry of plots
          1.2 enrichment of plot boundaries and nearby miombo with more
               careful choice of MPT's (indigenous and exotic)
          1.3 intensify/improve intercropping practices
          1.4 soil conservation techniques
          1.5 techniques to prevent leaching of nutrients
          1.6 cover crop and MPT mixtures to plant into Chitemene at
               end of cropping circle.

     2.   Modification of mound systems

          2.1 study size and spacing relative to plant residues,
               cropping cycle
          2.2 boundary or alley-cropping with fast-growing woody
               species to produce plant residues for mounds
          2.3 study decomposition/nutrient cycles for grass and woody
               plant residues (leaf, twig; indigenous and exotic plants)

     3.   Plant propagation trials

     4.   Home garden intensification/diversification

          4.1 integration of more fruit trees of higher value
               (commercial or home use)
          4.2 incorporate boundary or alley plantings (MPT/s) to
               improve crop yields and maintain permanent cultivation on
               plots
          4.3 diversify women's cassava plots ecologically and
               economically
          4.4 introduction and trials of leafy vegetable crops
               (indigenous and exotic)

     5.   Enrichment of Miombo woodland

Table 4. (Continued)

    5.1 replace existing species with more productive ones (wide-spread)

    5.2 increase the proportion of useful preferred species in fallows and in miombo areas near homesteads

6.   Screening of productive trees for dambos*

    6.1 screen MPT's for tolerance of waterlogging, vertisols

    6.2 test MPT's for dambo boundary production of residues for mounded vegetable plots.

---

*seasonally flooded depressions, with vertisols ("black cotton" soils)

users are to be incorporated as full partners in the research and development process.

To address the wide range of client needs and the widespread lack of ready-to-use information and agroforestry technologies requires simultaneous initiatives in four distinct types of research, all incorporating a common user perspective in the overall approach and the specific methodologies. The key elements of research content that require immediate attention are: selection and improvement of germplasm; improvement or development of prototype systems; adaptive research and action research.

6.3.1. Germplasm Selection and Improvement

Selection and improvement of germplasm for agroforestry systems should include non-woody as well as woody components. This process is currently fragmented and is mostly in the hands of foresters, some of whom now incorporate a multiple commodity approach (7, 8). What is actually needed is something much broader that taps rural peoples' knowledge meets new criteria set by rural consumers and assures continued access to local genetic resources and their derivatives (61, 71).

6.3.2. System Improvement or Development

Improvement or development of prototype systems for broad regional application is the current emphasis at ICRAF, IITA, ICRISAT, CATIE and other international research centers. However, most programs have gravitated toward a few prototype systems such as alley cropping for mulch. These technologies are usually tied to support of staple grain crops of special interest to the research organization or to urban consumers and national policy makers. More imaginative design and testing of prototypes is needed to meet the full range of land users' needs and to fit into the complex mosaics of rural landscapes. For example, the trees themselves may be the main crop, with shade tolerant leafy vegetables underneath. Living woody fences may provide support for fruit or vegetable vines. Leafy vegetables and medicinal plants may grow readily along the base of such fences. Leafy mulch for cropland may come from large dispersed trees, from the fenceline, or from a fenced block of trees near the home. The list of potential prototypes is long and varied and should be carefully considered by planners and clients prior to initiating detailed research on one or more designs (59, 6, 51).

6.3.3. Adaptive Research

Adaptive research can help to adjust prototypes to local conditions and to feed back new ideas to regional prototype research. Agroforestry programs within development "pilot projects" often constitute de facto adaptive and/or action research, whether intended as such or not. Ideally, adaptive research should result in specific changes to the prototype practice or to the receiving system, to fit the practice into local environments and land use systems. The changes might include new spacing, species, or management practices, all minor variations on the basic agroforestry design idea. Adaptive research should normally be based on a well known traditional practice or an experimentally derived system that fits the general conditions of land users in a given area. If the prototype is not appropriate, then the adaptive research will probably result, at best, in requests for a new start with a feasible prototype for local adaptation.

Adaptive research is currently carried out by isolated non-governmental organizations (NGO's) and national research projects. The prototype often consists of alley cropping as practiced at IITA, with the same species, regardless of user needs, preferences or ecozone. Some integration and upgrading of adaptive research efforts has been achieved by international NGO's, most notably CARE International (70, 6) and ENDA (27) in collaboration with national agencies (government and NGO).

6.3.4. Action Research

Action research tests and develops social, economic and policy mechanisms to enable successful design and application of agroforestry systems that fit rural peoples' needs. This process can give some local communities an active, controlling role in technology, management and policy experiments that result in local centers of demonstration, expertise (and possibly advocacy) serving the people in the region as well as the researchers and policy makers (11, 41). This type of research is usually conducted by isolated NGO's working at the grassroots level. Action research has been stronger in social forestry than in agroforestry, particularly in India (22), Nepal (9), the Philippines and at several sites in Latin America (74, 30).

How can a user perspective be applied to all four classes of agroforestry research? "Local participation", which may or may not be the same as a full user perspective, is usually limited to adaptive and action research. These are often not considered research at all, and thus fall to the users (and development community) by default, often mistakenly assigned to the domain of "extension" (61). However, technology adaptation and creation of appropriate conditions for community adoption or development of technology require more systematic research input.

For example, in a Farming Systems "extension" project on rotational grazing in Botswana, the field team introduced the same technology with the same management approach to 12 communities. From the beginning the project lacked social research input to define "communities" as units of operation, an issue that confounded efforts to understand people's response to the management scheme. The clients of the program rejected the technology and management package in 11 cases and adopted it readily in the 12th case (13). A social research component might have resulted in testing of alternative management schemes, or it might have addressed and explained the failure of the single design early on, with specific suggestions for changes in the technology design or management approach.

On the other hand, germplasm and technology prototype research are often reserved for the laboratory and the research station with occasional unacknowledged pilfering of users' ideas or knowledge. Moreover, the choice to emphasize one of these types of research in any given situation is often more a function of institutional bias than of the users' needs or aspirations. An appropriate mix of all four classes of research requires a prior diagnosis of the constraints and opportunities in specific land use systems.

The very separation of these classes of research from each other and from development and extension efforts is, in itself, a problem. The Zambia example (cited previously) illustrates the need for a combination of germplasm selection and improvement (indigenous and exotic), prototype technology development, adaptive research on existing technologies (mounding and home gardens), and action research (to explore compatible tenure, technology and landscape designs for diverse user groups under a variety of site and marketing conditions). However, four different institutions acting independently on each of these questions might well work at cross purposes to one another and might leave their clients with an unwieldy collection of incompatible tools and practices. The relative importance of each type of research also varies substantially from one case to another. In some cases this could easily be distorted by relative institutional strength in each of the four areas, if the four types of research are not integrated.

6.4. HOW CAN INSTITUTIONS DEVELOP AND APPLY A USER PERSPECTIVE?

Rather than to divide work equally, or on the basis of institutional strength, it may be more appropriate to mobilize expertise and concentration in some combination of the four classes of research around practically defined, shared goals, from the user's point of view (Table 5). To maintain a user focus in the content and method of such endeavors, matching expertise and concentration in the four aspects of a user approach should be woven into the research and action team as required by the work at hand.

Ideally, the very composition and mandate of such international and national research teams would be determined by prior consultation with clients to define the research and action objectives in users' terms and to identify the appropriate actors to get the job done. This suggests a centralized broker function at national or provincial level to identify (or respond to) representative areas and communities, to consult users and to help them to define their own priorities for decentralized research and action programs. Clients might also suggest specific local institutions and/or individuals to participate in the programs.

Where such an approach is in the realm of wishful thinking, it may still be possible to re-orient established, on-going projects and programs to users' concerns and to enrich their own capabilities for continuing innovation. Many development and research planners scoff at the ambition and impracticality of such an approach, citing the lack of trained personnel and lack of funds and/or institutional niches for social scientists on agroforestry teams. The options are in fact quite flexible.

Table 5. Research directions in users' terms.

Trees and shrubs for local or home subsistence use (as products/as services rendered)

Trees and shrubs as cash sources (direct/indirect)

Trees and shrubs as reserves, savings, inheritance

Trees and shrubs for increasing returns to labor (cash, products)

Fitting useful trees into "in-between" places, improving the spatial order of things

Facilitating and formalizing access to trees/shrubs and their products for resource-poor and landless people

---

Institutions can incorporate a user perspective into their work by hiring and/or training interdisciplinary people competent in relevant aspects of both biophysical and social arts and sciences (multipurpose people). Some institutions can more readily form multidisciplinary teams from previously separate departments of research and extension or socioeconomic and biophysical research. It should be noted, however, that not all "social scientists" can or will take on such a role. This is in fact an art of listening, translating between worldviews and applying social understanding to creative technology design.

In cases with no such expertise in house and no scope for expansion of staff, it may still be feasible to contract short term user-focus consultants to help determine research and action priorities, and to train regular staff in participatory research and action methods. Yet another possibility is a partnership between research and development agencies, or between institutions experienced in technology and social innovations, respectively. In some cases, agroforestry institutions might team up with other technical research and action agencies which have relevant experience and expertise in a user-focused approach, whether it be in primary health care, rural water supply, literacy training or some other rural service. Collaboration of agroforestry institutions with other centers of expertise and innovation might also extend to community development catalysts linked in a far-flung decentralized network.

Organizational and institutional models of all types have sprung up in response to the opportunities and needs for agroforestry research and action. Numerous methodologies have also been developed, tested and adapted for involvement of clients (to varying degrees) in the process of technology development, adaptation and management (Table 6) within a variety of institutional settings. The question is not whether agroforestry research and action programs can incorporate a user perspective, but rather how they will do so if they are to fulfill their obligations to their clients.

Table 6. Reported practices indicative of a partial user perspective in several agroforestry research and rural development institutions. + = present; ++ = heavily utilized; +++ = well developed; blank = no positive report found. Reports indicate specific projects or researchers, not necessarily institutional practices.

	Use rapid appraisal system	On-farm research and demonstration	Treat multiple uses	Stratify user groups	Treat land tenure/access issues	Treat whole landscapes	Involve communities	Incorporate indigenous knowledge	Involve clients as researchers
ICRAF[1]	+++	++	+++	+	+	++	+	+	+
IITA[2]/IICA[3]	+	++	+	+					+
ICRISAT[4]	+		+						
Beijer Inst/KWDP[5]	+	+++		++	+	+	++	++	++
CARE Int'l[6]	++	+++	++	+			+++	+	+
ENDA[7]	+	++	++				+++	++	+
ICAR[8]	++	+	++	+		+			
KEFRI[9]	++	++	++			+	+	+	+
Wageningen Univ.[10]	+++	+	+++	+	++	+++		+	

[1] International Council for Research in Agroforestry, Nairobi, Kenya;
[2] Iternational Institute for Tropical Agriculture, Ibadan,Nigeria;
[3] International Livestock Centre for Africa, sub-centre at IITA, Nigeria;
[4] International Centre for Research in the Semi-Arid Tropics, Hyderabad, India;
[5] Beijer Institute/Kenya Wood fuel Development Project, Nairobi, Kenya;
[6] CARE International, Agriculture and Natural Resources, New York;
[7] Environment and Development Action, Dakar, Senegal (Zimbabwe, Caribe);
[8] Indian Council for Agricultural Research;
[9] Kenya Forestry Research Institute;
[10] Wageningen Agricultural University, Dept. of Landscape Architecture (The Netherlands)

REFERENCES

1. Berry, S., Geyer, J., Peters, P., Issacman, A., Moore, A., Richard, A. and M. Watts. 1983. The food crisis and agrarian change in Africa. Social Sci Res Center, Nat Res Council, Washington, D.C.
2. Bhatt, C. P. 1982. Ecosystem of the central Himalayas and Chipko movement, determination of hill people to save their forests. In: Dashauli Gram Swarajya Sangh, pp 7-31. Gopeshwar, Uttar Pradesh, India.
3. Biggs, S. D. 1980. Informal R & D. Ceres 13:23-6.
4. Brokensha, D., Warren, D. M. and O. Werner (eds). 1980. Indigenous knowledge systems and development. Lanham, Univ Press Amer.
5. Buch, N. 1980. Role of women in nutrition gardens in social forestry. In: Proc Seminar Role Women Commun For, Dec 4-9, 1980. For Res Inst Coll, Dehra Dun, India.
6. Buck, L. 1986. Training manual for extension workers. CARE Intern, Nairobi, Kenya.
7. Burley, J. F. 1985. Global needs and problems of the collection, storage and distribution of MPT germplasm. Intern Council Res Agrofor (ICRAF), Nairobi, Kenya.
8. Burley, J. F. and P. Carlowitz (eds). 1984. Multipurpose tree germplasm. Intern Council Res Agrofor (ICRAF), Nairobi, Kenya.
9. Campbell, J. G. and T. N. Bhattrai. 1983. People and forests in hill Nepal. Proj Paper No 10. HMG/UNDP/FAO, Commun For Dev Proj, Kathmandu, Nepal.
10. Cernea, M. 1985. Units of social organization sustaining alternative forestry development strategies. Agric Rural Dev Dep, The World Bank. Washington, DC.
11. Chambers, R. 1983. Rural development: putting the last first. Longman, London, UK.
12. Chambers, R. 1984. To the hands of the poor: water, trees and land. Disc Paper No 14, Ford Foundation, New Delhi, India.
13. Chandler, D. 1983. Grazing systems in communal areas. In: Animal Prod Res Unit, Ministry Agric, Livestock and range research in Botswana 1981-1982, pp 29-38. Gaborone, Botswana.
14. Clay, J. 1983. Draft bibliography of indigenous agroforestry systems. Cultural Survival Quarterly, Cambridge, MA (mimeo).
15. Das Gupta, M. 1984. Microperspectives on the slow rate of urbanization in India-informal security system and population retention in rural India. Sem micro-Approaches Demographic Res, Sept 1984. Canberra, Australia (presented paper).
16. Doherty, V. S., Miranda, S. M. and J. Kapen. 1982. Social organization and small watershed management. In: The role of anthropologists and other social scientists in interdisciplinary teams developing improved food production technology. Intern Rice Res Inst (IRRI), Los Banos, Philippines.
17. Duchard, I. 1986. Landscape planning and agroforestry. Ann Conf, Council Educ in Landscape Architecture, Sept 18, 1986. Athens, Georgia.
18. FAO. 1986. Tree growing by rural people. UN/Food Agric Org (FAO), Rome, Italy.
19. Fernandes, E. C. M. and P. K. R. Nair. 1986. An evaluation of the structure and function of tropical homegardens. Working Paper No 38, Intern Council Res Agrofor (ICRAF), Nairobi, Kenya.

20. Flora, C. B. 1984. Intra-household dynamics in farming systems
    research: the basis of whole farm monitoring of farming systems
    research and extension. Farming Syst Support Proj, Kansas State Univ,
    Manhattan, KA.
21. Flores Paitan, S. 1987. Investigacion agroforestal en la Cuenca
    Amazonica. Intern Council Res Agrofor, (ICRAF), Nairobi, Kenya
    (draft).
22. Foley, G. and G. W. Barnard. 1985. Farm and community forestry.
    Earthscan Energy Infor Prog, London, UK.
23. Fortmann, L. 1985. The tree tenure factor in agroforestry with
    particular reference to Africa. Agrofor Syst 2:229-251.
24. Fortmann, L. and D. Rocheleau. 1984. Women and agroforestry: four
    myths and three case studies. Agrofor Syst 2:253-272.
25. Freire, P. 1970. Pedagogy of the oppressed. Seabury Press, New York,
    NY.
26. Geertz, C. 1972. The wet and the dry: traditional irrigation in Bali
    and Morocco. Human Ecol 1:23-39.
27. Geilfus, F. 1986. Un projecto de accion cumunitaria y de
    investigacion de campo con pequenos agricultores. Envir Develop
    Action (ENDA-Caribe), Santo Domingo, Dominican Republic.
28. Haug, R. 1981. Agricultural crops and cultivation methods in the
    Northern Province of Zambia. Occ Pap 1, Dep Agric Econ, Agric Univ,
    As, Norway.
29. Haugerud, A. 1986. An anthropologist in an african research
    institute: an informal essay. Bull Inst Develop Anthropol. 4:2.
30. Herran, J. 1985. An experience in fieldwork. Global Meeting Environ
    Develop, February 4-8, 1985. Environmental Liaison Centre, Nairobi,
    Kenya (presented paper).
31. Holden, S. T. 1986. The role of caterpillars in woodland management,
    diet and local economy in Northeast Zambia. Agric Univ, As, Norway
    (mimeo).
32. Hoskins, M. 1982. Social forestry in West Africa: myths and
    realities. Ann Meeting, Amer Assoc Adv Sci (AAAS), Washington, DC
    (presented paper).
33. Hoskins, M. 1983. Rural women, forest outputs and forestry projects.
    UN/Food Agric Org (FAO), Rome, Italy.
34. Huxley, P. A., Rocheleau, D. E. and P. Wood. 1986. Farm systems and
    agroforestry research in north Zambia: Phase I report-Diagnosis of
    land use problems and research indications. Intern Council Res
    Agrofor (ICRAF), Nairobi, Kenya.
35. Jain. 1984. Standing up for the trees: Women's role in the Chipko
    movement. Unasylva 36:146.
36. Jiggins, J. 1986. Problems of understanding and communication at the
    interface of knowledge systems. In: Poats, S., Spring, A. and M.
    Schmink (eds) Gender issues in farming systems research and
    extension. Westview Press, Boulder, CO.
37. Joshi, C. 1982. Men propose, women oppose the destruction of forests.
    Infor Serv Sci Society-Related Issues, Centre Sci Environ, New Delhi,
    India.
38. Lavalle-Legaspie, C. 1985. Summary of work done by Promocion de
    Desarrollo Popular, Mexico. Global Meeting Environ Develop, Feb 1985.
    Environment Liaison Centre, Nairobi, Kenya.
39. Mansfield, J. E. 1973. Summary of research findings in Northern
    Province Zambia. Suppl Rep No 7, Land Resource Div, Ministry Overseas
    Develop, London, UK.

40. Mattson, L. 1986. Report of agroforestry diagnostic exercise. Chief Mwamba area, Kasana, Zambia. In: Huxley, P. A., Rocheleau, D. E. and P. Wood (eds), Farm systems and agroforestry research in north Zambia: Phase I report - Diagnosis of land use problems and research indications. Intern Council Res Agrofor, (ICRAF), Nairobi, Kenya.
41. Max-Neef, A. 1982. From the outside looking in: experiences in barefoot economics. Dag Hammarskjold Foundation, Uppsala, Sweden.
42. Maydell von, H. J. 1979. Agroforestry to combat desertification: A case study of the Sahel. In: Agroforestry: Proceedings of the 50th Symp Trop Agric. Bull 303, Dep of Agric Res, Koninklijik Inst voor de Tropen, Amsterdam, Netherlands.
43. Michon, G. 1983. Village-forest-gardens in West Java. In: P. Huxley (ed) Plant research and agroforestry, pp 13-24. Intern Council Res Agrofor (ICRAF), Nairobi, Kenya.
44. Mishra, A. and S. Tripathi. 1978. Chipko Movement, Uttarakhand women's bid to save forest wealth. People's Action, New Delhi, India.
45. Muntemba, S. 1982. Women as food producers and suppliers in the twentieth century: the case of Zambia. Develop Dialogue 1-2:29-50.
46. Nair, P. K. R. 1987. Agroforestry systems in major ecological zones of the tropics and subtropics. Working Pap No 47, Intern Council Res Agrofor (ICRAF), Nairobi, Kenya.
47. Nations, J. and D. Komer. 1982. Indians, immigrants and beef exports: deforestation in Central America. Cultural Survival Quarterly 6:8-12.
48. Okafor, J. C. 1980. Edible indigenous woody plants in the rural economy of the Nigerian forest zone. For Ecol Manage 3:45-55.
49. Okafor, J. C. 1981. Woody plants of nutritional importance in traditional farming systems of the Nigerian humid tropics. PhD Dissert, Dep For Resourc Manage, Univ Ibadan, Ibadan, Nigeria. 369 pp.
50. Opole, M. 1985. With culture in mind: improved cookstoves and Kenyan women. In: Munyakho, D. and R. Kurtz (eds) Women and the environmental crisis. Proc Workshop Women Environ Develop, Non-governmental organization (NGO) women's forum, Environment Liaison Centre, July 10-20, 1985. Nairobi, Kenya.
51. Raintree, J. B. 1983. A diabnostic approach to agroforestry design. Intern Symp Strategies Designs Afforest, Reforest Tree Planting. Hinkeloord, Wageningen, Netherlands (submitted).
52. Raintree, J. B. 1987. Agroforestry, tropical land use and tenure. In: J. B. Raintree (ed) Land, trees and tenure. Proc Intern Workshop Tenure Issues Agrofor. Intern Council Res Agrofor (ICRAF), Nairobi, Kenya.
53. Raintree, J. B., Rocheleau, D., Huxley, P., Wood, P. and F. Torres. 1984. Report, Joint ICAR/ICRAF diagnostic and design exercise at the Bhaintan Watershed in the Outer Himalaya of Uttar Pradesh. Intern Council Res Agrofor (ICRAF), Nairobi, Kenya (mimeo).
54. Rhoades, R. and P. Booth. 1982. Farmer-back-to-farmer: a model for generating acceptable agricultural technology. Agric Admin 11:127-137.
55. Richards, P. 1985. Indigenous agricultural revolution. Hutchinson and Co, Ltd. London, UK. 192 pp.
56. Rocheleau, D. 1983. Ecosystem analysis in D & D applications. In: J. B. Raintree (ed) Resources for agroforestry diagnosis and design, pp 137-155. Working Pap No 7. Intern Council Res Agrofor (ICRAF), Nairobi, Kenya.

57. Rocheleau, D. 1984a. An ecological analysis of soil and water conservation in hillslope farming systems: Plan Sierra, Dominican Republic. PhD Dissert, Univ Florida, Gainesville, FL.
58. Rocheleau, D. 1984b. Prog Rep. ICRAF collaboration in the CARE-Kenya Siaya agroforestry project. Intern Council Res Agrofor (ICRAF), Nairobi, Kenya.
59. Rocheleau, D. 1985. Criteria for re-appraisal and re-design: intra-household and between-household aspects of FSRE in three Kenyan agroforestry projects. In: Flora, C. and M. Tomacek (eds) Selected proceedings. Ann Symp Farming Syst Res Extension, Oct 7-18, 1984. Kansas State Univ, Manhattan, KA.
60. Rocheleau, D. and A. van den Hoek. 1984. The application of ecosystems and landscape analysis in agroforestry diagnosis and design: a case study from Kathama Sublocation, Machakos District, Kenya. Work Pap No 11, Intern Council Res Agrofor (ICRAF), Nairobi, Kenya.
61. Rocheleau, D. E. and J. E. Raintree. 1986. Agroforestry and the future of food production in developing countries. Impact Sci Soc 142:127-141.
62. Stollen, K. A. 1983. Peasants and agricultural change in northern Zambia. Occ Pap 4, Dep Agric Econ, Agric Univ Norway, As, Norway.
63. Sutherland, A. J. 1987. The benefits of adopting a community approach to farmer selection. Workshop Household Issues Farming Syst Res, April 27-30, 1987. CIMMYT, Lusaka, Harare, Zimbabwe (presented paper).
64. Teel, W. 1985. A pocket directory of trees and seeds in Kenya. Kenya energy non-governmental org, Nairobi, Kenya. 151 pp.
65. Thompson, J. 1983. Participation, local organization, land and tree tenure: future directions for Sahelian forestry. CILSS/Club du Sahel, Ouagadougou/Paris, France. 34 pp.
66. Tibaijuka, A. K. 1984. An economic analysis of smallholder banana-coffee farms in the Kagera Region, Tanzania: causes of decline in productivity and strategies for revitalization. PhD Dissert, Agric Univ, Uppsala, Sweden. 250 pp.
67. Tinker, I. 1976. The adverse impact of development on women. In: Tinker, I. and M. B. Bramsen (eds) Women and world development, pp 23-34. Praeger Publishers, New York, NY.
68. Vedeld, Y. 1981. Social-economic and ecological constraints on increased productivity among large circle chitemene cultivation in Zambia. Occ Pap 2, Dep Agric Econ, Agric Univ, As, Norway.
69. Vonk, R. 1985. Annual Report, CARE/Kenya agrofor proj. CARE International, New York, NY.
70. Vonk, R. 1986. Annual Report, CARE/Kenya agrofor proj, CARE International, New York, NY.
71. Wachira, K. 1986. The state of seeds in Kenya. Kenya Energy Non-governmental Org, Nairobi, Kenya.
72. Weber, F. 1986. Reforestation in arid lands. VITA. Arlington, VA.
73. Weber, F. and M. Hoskins. 1983. Agroforestry in the Sahel. Rep Dep Sociol, Virginia Polytech Inst State Univ, Blacksburg, VA.
74. Wiff, M. 1984. Honduras: women make a start in agroforestry. Unasylava 36:146.
75. Woods, B. 1983. Altering the present paradigm: a different path to sustainable development in the rural sector. The World Bank, New York, NY.

7. PLANT-INSECT INTERACTIONS AND SOIL FERTILITY RELATIONS IN AGROFORESTRY SYSTEMS: IMPLICATIONS FOR THE DESIGN OF SUSTAINABLE AGROECOSYSTEMS

Miguel A. Altieri, F. Javier Trujillo, and John Farrell*

Division of Biological Control, University of California, Berkeley, CA 94720

*Agroecology Program, University of California, Santa Cruz, CA 95064

7.1. ABSTRACT

The development of sustainable agroecosystems requires an agroecological approach, sensitive to the socio-economic and agronomic processes that govern agricultural production. The goal is to design production systems tailored to the subsistance and income needs of local people and the resource base of regional agroecosystems. Understanding the complexities and dynamics of traditional farming systems through agroecological studies can provide important guidelines to direct this development strategy. Rural peoples' knowledge may be blended with modern science to improve progressively on the productivity of the proposed systems. Research results from our studies in traditional agroforestry systems in Tlaxcala, Mexico and diversified apple orchards in northern California are used to illustrate the above points.

7.2. INTRODUCTION

In the last two decades many scientists have started to apply ecological principles to study, design and manage agricultural systems (20). Two important approaches stand out among agronomists and ecologists. First, agricultural systems are regarded as experimental units on which ecological processes and hypotheses can be studied and tested (i.e., studies on predator/prey relationships, plant-plant and plant-insect interactions, nutrient cycling, etc.; 30). Second, ecological principles governing fundamental properties of stability, productivity, soil fertility, population regulation, etc., are considered for the design of management practices leading to optimal performance of agroecosystems (3). Both approaches aim at recapturing for agroecosystems the protective functions and stability that characterize natural ecosystems. The idea is to develop an agroecosystem that emulates later stages of natural succession as much as possible, for this is how, it is claimed, biological stability can be achieved. In Costa Rica, Ewel et al. (25) tested this hypothesis using the local natural vegetation as an architectural and botanical model for designing locally adapted agroecosystems. They conducted spatial and temporal replacements of wild species by structurally similar cultivars, maintaining through time a continual cover, thus avoiding site degradation and nutrient leaching, and providing crop yields throughout the year.

Social scientists, on the other hand, have mostly focused on the interactions between social systems and ecological systems mediated by the

exchange of energy, materials and information (24). A major tendency of
anthropologists, geographers and rural sociologists has been to describe
and analyze the different patterns of human subsistence in traditional
agricultural communities (35, 15). Important questions examined have
dealt with the social relations of production, the interactions between
people and the environment as well as interactions of villages and the
outside world (36). Much of this work has contributed to build a much
needed human ecology perspective to agroecosystem research (34). Although
assessments of traditional 'know how' and anthropological and
ethnobotanical studies of traditional farming systems in the Third World
are numerous (29, 11), they have mainly focused on the patterns of
behavior followed by an ethnographic unit in the realm of agricultural
technology, which results in typical sets of land utilization in time and
space, seasonal distribution of labor, nutrition and other needs, rarely
examining the ecological mechanisms that underline the sustainability of
these agroecosystems.

Traditional farming systems have emerged over centuries of cultural
and biological evolution and represent accumulated experiences of
interacting with the environment by farmers without access to external
inputs, capital, or scientific knowledge. Such experience has guided
farmers in many areas to develop sustainable agroecosystems, managed with
locally available resources and with human/animal energy (7). Most
traditional agroecosystems are based on the cultivation of a diversity of
crops in time and space allowing farmers to maximize harvest security
under low levels of technology (17, 18). Because these systems still
utilize minimal inputs, lack continuous disturbances and exhibit complex
interactions among and between people, crops, soils, animals, etc., many
agroecologists regard them as unique settings to evaluate properties of
stability and sustainability (27).

The benefits derived from studying these systems are two-fold.
First, as change occurs in the face of inevitable agricultural
modernization, knowledge of the traditional cropping patterns and
management practices and ecological rationale behind them are being
gradually lost. Because modern agricultural development, characterized by
broad-scale technological recommendations, has largely ignored the
environmental, cultural and socio-economic heterogeneity characteristic of
traditional agriculture, there has been an inevitable mismatching of
agricultural development and the needs and potentials of local people and
localities (10, 19). By understanding the features of traditional
agriculture, such as the ability to bear risk, and the production
efficiencies of symbiotic crop mixtures, etc., it is possible to obtain
important information, which may be used for developing appropriate
agricultural strategies more sensitive to the complexities of
agroecological and socio-economic processes. The design of farming
systems and techniques tailored to the needs of specific peasant groups
and regional agroecosystems must also be based on elements of rural
people's knowledge properly rescued in the course of modernization
(16).

Second, ecological principles extractable from the study of
traditional agroecosystems can be used to design new, improved,
sustainable agroecosystems in the developed countries and thus correct the
many deficiencies affecting modern agricultural (3). Modern agricultural
systems are a product of structural evolution that substitute high energy
inputs for stabilizing ecological interactions. Many significant
ecological interactions of interest to develop alternative production

systems simply are not present in current, heavily disturbed monocultures. Agricultural scientists searching for designs of alternative systems can find in areas of traditional agriculture, sites that do not have as their goal maximum net crop yields, but interaction between as many biological components as possible (23, 4).

As long as modern agricultural policies emphasize greater yields, the ecology-farming linkage will often be broken as ecological principles are ignored or overriden. Today, these breakdowns manifest themselves as recurrent pest outbreaks in many cropping systems and also in the form of salinization, soil erosion, pollution of water systems, etc. Alternatively, agricultural strategies based on diversified farming systems (i.e., agroforestry systems, polycultures, etc.) can lead to achievement of moderate to high levels of productivity by manipulating and exploiting resources that are internal to the farm. The resulting systems are more sustainable and economical, thus increasing the equity of the system. Thus, new strategies must broaden their performance criteria to include properties of sustainability, stability and equity along with the goal of increased production.

In our search for elements of sustainability in agroecosystems, we have studied a number of agroecosystems both traditional and modern, which contain a wealth of information on efficient crop production under varying resource, biological and socioeconomic constraints (5, 6, 8). In this paper we describe results from studies conducted in traditional corn production systems in Tlaxcala, Mexico and in organic apple orchard systems in northern California, which exhibit damped oscillations of insect pest populations and soil fertility restoring properties, both important ecological functions leading to relatively stable agricultural yields. We discuss the implications of these research results for the design of agroforestry systems.

7.3. THE AGROECOLOGY OF CORN PRODUCTION IN TLAXCALA

7.3.1. Cropping systems

The agriculture of the state of Tlaxcala, Mexico, has a rich history dating to prehispanic times (13, 38). Tlaxcalan farmers have continuously faced marginal conditions of natural resources, low soil fertility and extreme climatic limitations. They have, however, managed to survive in this fragile environment, through a process of ecological modification and adaptation based on diversified farming, genetic diversity maintenance, and unique land and water management practices. Corn production is the foundation of Tlaxcalan agriculture, and this production is safeguarded by growing corn mixed with annual and perennial plants in polycultural and/or agroforestry patterns, and also in genetically complex fields that maintain a mosaic of traditional varieties over time and space (39). This production system is closely related to the traditional Tequexquinahuac (28). Trees are conspicuous components of Tlaxcala agroecosystems and a number of species (i.e., capulin (_Prunus_ sp.), sabino (_Juniperus_ sp.), tejocote (_Crataegus mexicana_), peach (_Prunus persica_), apple (_Pyrus malus_) and tepozan (_Buddleja americana_)) are integrated with corn in a variety of patterns resembling a veritable crop savannah. These trees are not commonly planted by farmers but are maintained along field borders or left scattered within fields with corn planted right up to their base. Still higher tree densities can be found where narrow fields have been cleared and planted to crops within dense woodlands. Trees and corn are also mixed in orchards, and in the fruit-growing areas apple, peach, plum, pear-apple, apricot and walnut are commonly interplanted with corn and alfalfa (26).

Also common are mixtures of corn and frijol (_Phaseolus_ spp.) or corn and fava beans (_Vicia fava_). The planting of corn with frijol is traditionally done by sowing both plants seeds in the same hole, whereas fava beans are normally planted between mounds of corn within a given row. Both types of beans are important food sources and are also recognized for their soil improving qualities ("buen rastrojo"). The corn-fava bean combination provides the additional benefit of risk aversion in areas prone to frost damage, since fava beans are fairly tolerant to frost conditions.

The rotation of corn and alfalfa is very common in areas with available irrigation or with deep soils characterized by sufficiently high moisture-holding capacity. Alfalfa is grown as animal fodder and planted with corn in a strip-cropping arrangement that changes every year, as corn is grown where alfalfa grew the previous year and vice-versa. Other crops often grown in rotation with corn are wheat and barley, which are valued for their suppressive effect on certain weed species. There are numerous variations in rotation patterns, including corn-barley (or wheat)-corn, corn-fava bean (or frijol)-corn and corn-fava bean-barley-corn.

The Tlaxcalan agroecosystems differ from their Mexican traditional models in a number of ways. Notably, corn is grown under a number of temporal (rotations) and/or spatial arrangements (intercropping, strip-cropping, agroforestry) instead of in a cyclic polyculture of agroforestry pattern characteristically used by farmers in the lowland tropics (12). Tlaxcalan systems are genetically diverse, and this diversity results from interactions far more complex than the random planting of numerous corn varieties. Most notable, however, is the multiple use orientation of the systems; they serve as sources for local household needs and also provide cash returns from sale of the various products.

These agroforestry and polyculture systems provide a unique opportunity to examine the influence of vegetational diversity of plant-soil relationships and on the population levels of insect pests and associated natural enemies, both important ecological processes that actively interplay within these mixed farming systems.

7.3.2. Tree Influences on Soil Properties

Surface soil samples to a depth of 15 cm were collected at 2, 6, 10, 14, 18 and 22 meters from the base of capulin and sabino trees along four randomly chosen transects in four traditionally managed corn fields (26). A decreasing gradient in all soil properties measured was observed with increasing distance from the trees. Soil nitrogen, phosphorus, CEC and calcium gradients were especially pronounced in the corn-capulin systems, suggesting that capulin had a greater influence on soil properties than sabino. Available phosphorus increased 4-7 fold and nitrogen 1.5-3 fold under the trees (Figure 1a and b). Total carbon, calcium, magnesium and CEC also increased. Potassium levels also descreased with distance from the trees, although its concentration was higher under sabino canopies (Figure 1c). Soil pH and percent moisture were also slightly greater under tree canopies. Both trees appeared to influence surface soil properties within a radius of 6-10 meters from their base.

92

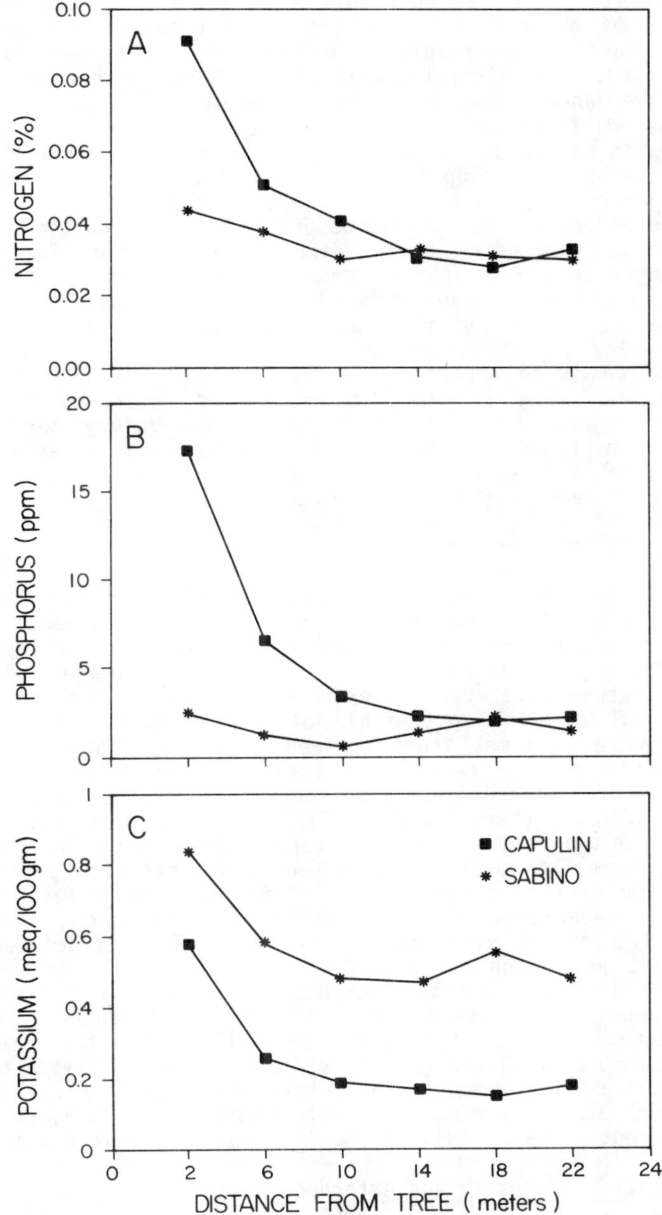

Figure 1. Change in major surface soil nutrient levels in a Tlaxcalan corn-woodland agroforestry system, with increasing distance from individual capulin (_Prunus capuli_) and sabino (_Juniperus_ spp.) trees. A = Nitrogen, B = Phosphorus and C = Potassium.

The redistribution of nutrients with litterfall and the resulting accumulation of organic matter under the trees are probably the only significant factors determining the observed trends of soil chemical properties (40). The higher concentration of carbon beneath capulin and sabino followed an expected trend due to the accumulation of litter by the trees. The distribution of nitrogen under the trees also seemed correlated with carbon accumulation. This is not surprising since most of the soil nitrogen is tied up in organic matter and should be higher in the zone of organic matter accumulation (14). Accumulation of exchangeable cations also appears to be closely associated with organic matter.

Phosphorus markedly accumulated beneath the capulin trees, corresponding with the phosphorus amounts deposited with leaf litter. In addition, the higher pH found beneath the trees compared to in the open seemed to account for the increase in phosphorus availability in that area.

The increased capacity of the soils under capulin and sabino to retain water was due, in part, to the higher levels of organic matter beneath the trees. More important to the understory corn than total amount of water present in the soil, is the amount of water available for use (measurable difference in moisture between 1/3 and 15 bars) which was somewhat higher beneath the trees than away from them.

7.3.3. Corn Yielding Performance

Corn yields varied considerably among agroforestry systems (Table 1). However, since the data proceeds from non-replicated farmers' fields, the effects of trees on corn yields cannot be statistically separated from other sources of variation, such as soil, microclimate and management differences between systems. Nevertheless, the data provide a tentative description of the yield potentials of each system. Substantial yield differences were at times found between corn plants growing adjacent to trees and away from the trees in the center of the fields, although these differences varied according to tree species and systems involved.

Yields of early (March) and late (May) planted maize were measured in three zones in two sabino woodland fields. Zone 1 was located along the field borders, while zones 2 and 3 were progressively farther from the trees into the center of the fields. There were no differences in final grain yield between zones 2 and 3, while maize in zone 1 produced about 33% less. Significant yield differences were observed between the field border (270 ± 95.1) and center (573 ± 90.9) zones in the late-planted fields. No significant yield differences were found between maize growing along the edge (yield index = 729 ± 102) or in the center (734 ± 279) of fields bordered by rows of maguey. Similarly, no differences in yield were found among corn plants grown next to a border of ailite trees (491 ± 88.3) and corn plants grown in the center rows (554 ± ± 116) of a field. Yields of corn plants growing near apple trees (516 ± 59.3) or five meters away in the center (597 ± 191) of the interrow strips were similar.

7.3.4. Vegetation Diversity and Arthropod Dynamics

From May through August 1983, soil dwelling and foliage insect communities were weekly monitored in five corn fields of varying degrees of vegetational diversity.

The vegetational arrangements under which corn was found affected the arthropod fauna. Some systems exhibited particularly high populations of

Table 1. Average yield, height and density of maize growing in association with various perennial plants in agroforestry systems (Tlaxcala, Mexico, 1983).

System	Yield[1]	Height (cm)	Density (No. plants/ha)
Apple orchard	597	201	34,515
Pear-apple orchard	912	245	33,638
Sabino woodland			
early-planted	1127	212	51,772
late-planted	573	173	40,950
Ailite woodland	617	247	32,175
Maguey	734	162	45,922
NI (capulin)[2]	1005	192	57,623
NI (sabino)	1073	191	40,950

[1]Yield expressed as a relative index per 4-meter transect.

[2]Yields from the non-influence (NI) zone around capulin and sabino are included for comparison and are gegarded as equivalent to monoculture yields.

corn feeding insects. For example, the scarab pest, _Macrodactylus_ sp. was predominantly more abundant in the corn-alfalfa strip cropping system, whereas in the corn system growing adjacent to ailite woodlands it reached the lowest densities (Figure 2a).

The predator _Orius_ sp. (Hemiptera: Anthocoridae) was more abundant in the corn-orchard and corn-alfalfa systems than in other systems (Figure 2b). The only system that exhibited high densities of several entomophagous insects was the corn-fava bean-squash polyculture. Populations of the coccinellid beetles _Hippodamia convergens_ and _Coccinella nugatoria_, and of a crab spider species (Thomisiidae) were higher in this polyculture system than in corresponding monocultures (Figure 2c).

The proximity of alfalfa strips to corn had a fundamental effect on the occurrence of all arthropods. As observed in Figure (3), coccinellids, _Collops_ (Coleoptera: Malachiidae) and _Orius_ were seasonally more abundant (about 30%) on corn plants located in the row adjacent to alfalfa than in the row in the center of the field, away from the alfalfa strip. _Macrodactylus_, however, showed the opposite trend. In systems where alfalfa was periodically cut, we did not observe these abundance

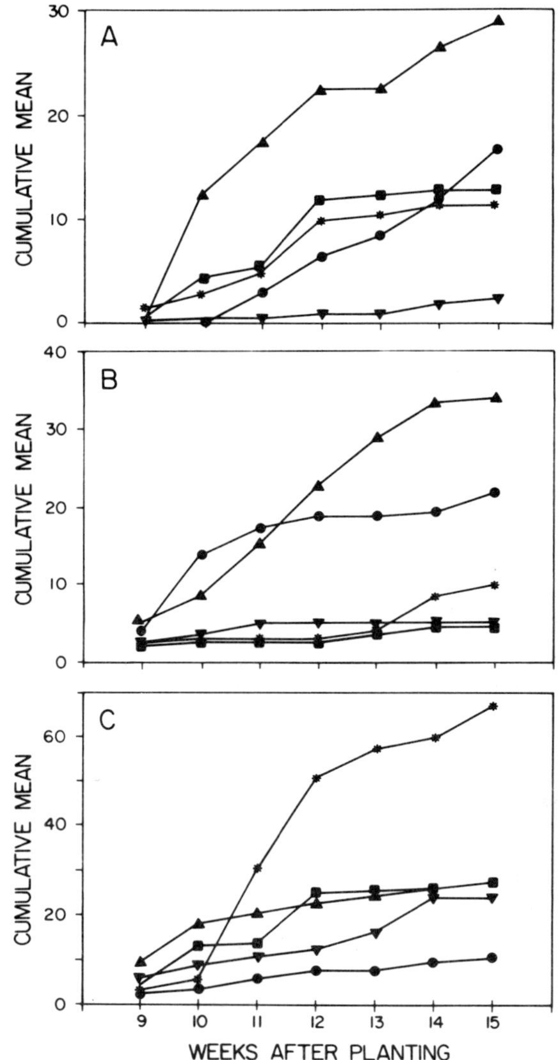

Figure 2. Cumulative mean densities per 100 corn plants of the pestiferous scarab beetle _Macrodactylus_ sp.(A), the hemipteran predator _Orius_ sp. (B), the lady beetles _Hippodamia convergens_ and _Coccinella nugatoria_ (C), in a range of corn cropping systems in Tlaxcala. ■ = corn monoculture, * = corn-fava bean intercropping, ▲ = corn-alfalfa strip cropping, ▼ = agroforestry system of corn under a tree layer of sabino and ailite trees, and ● = corn intercropped within an apple orchard.

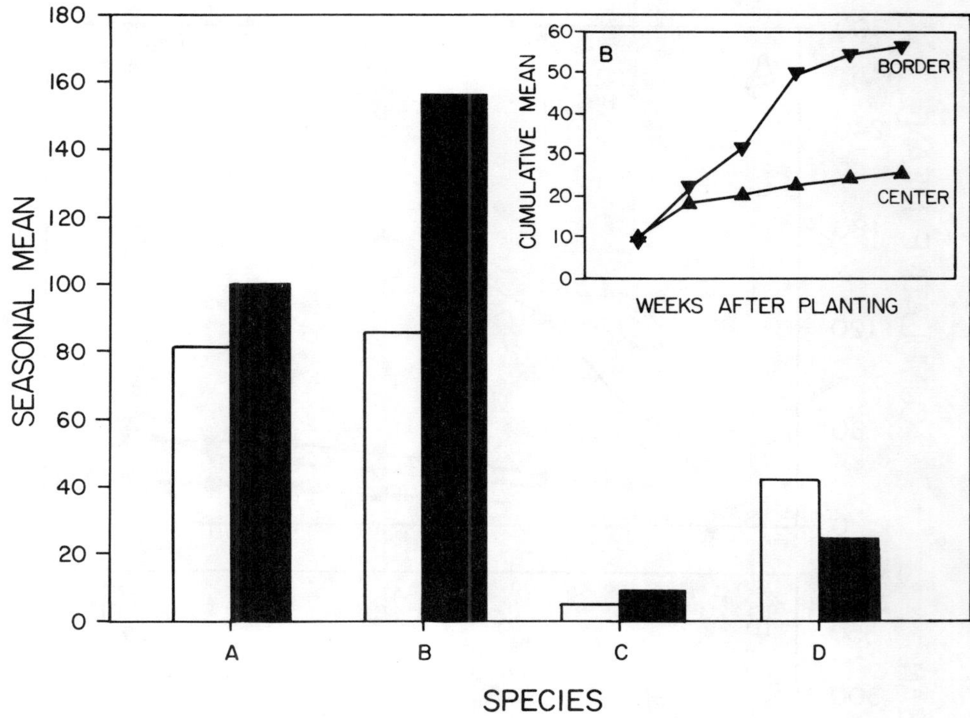

Figure 3. Seasonal abundances of predaceous <u>Orius</u> (A), Coccinellidae (B), <u>Collops malachiid</u> beetles (C) and the pest, <u>Macrodactylus</u> (D), on corn plants located in rows immediately adjacent to alfalfa strips and in rows located in the center of the field away from the alfalfa in Tlaxcala. The line graph in the upper right corner illustrates the seasonal abundance patterns of Coccinellidae in the border and center rows of corn.

gradients. Cutting forces arthropods to disperse, and through redistribution they attain similar densities throughout the corn fields.

Pitfall catches yielded significantly higher numbers of lycosid spiders in the corn-alfalfa system than in the other corn systems (Figure 4a). Ground beetles (Coleoptera: Carabidae) reached highest densities in the corn-fava bean-squash polyculture, apparently encouraged by the shelter provided by the squash leaves. Similarly, as in the case of foliage predators, substantially more lycosid spiders were caught in pitfalls placed in corn rows adjacent to alfalfa than in the center rows of the corn field (Figure 4b).

7.3.5. Design Implications

Growing corn in a multiple array of vegetation designs is a strategy well adapted to the plethora of microenvironments to which corn production

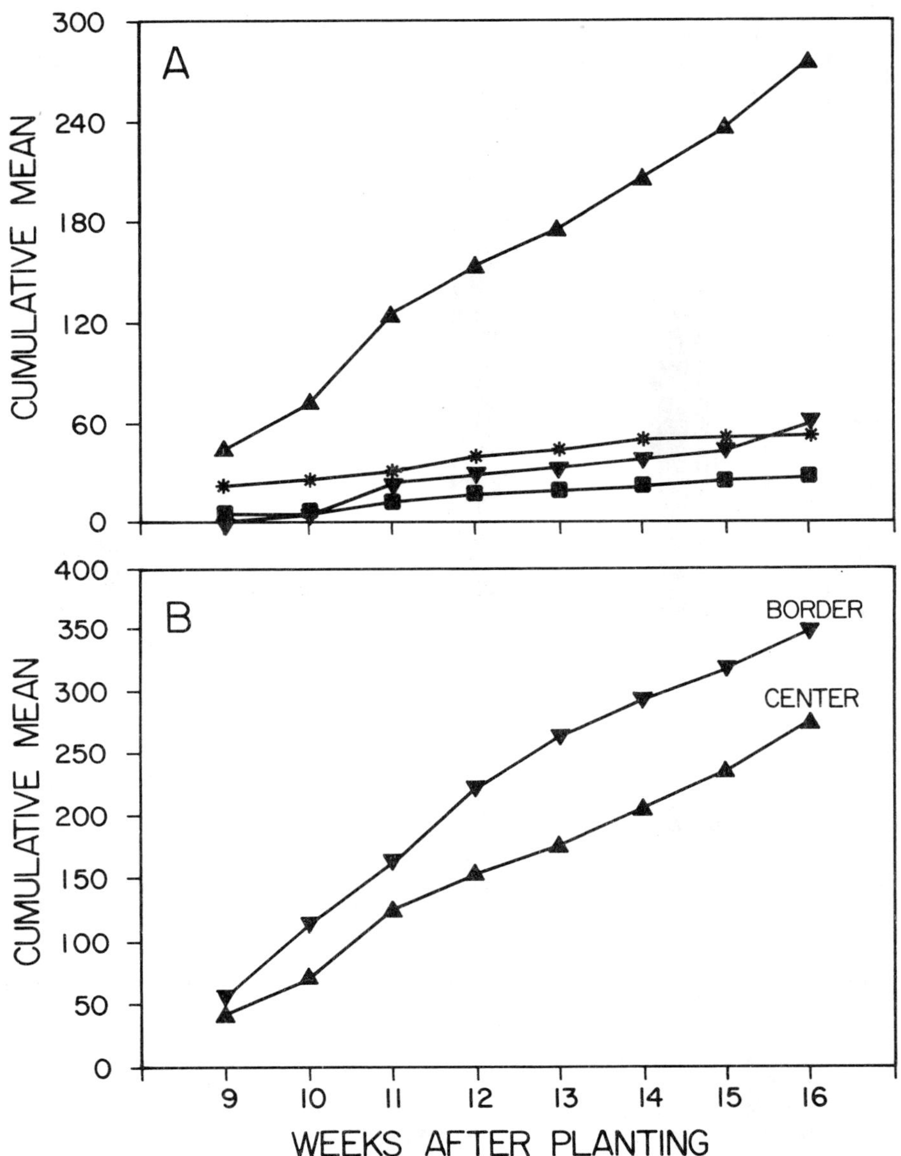

Figure 4. A. Numbers of lycosid spiders caught in pitfall traps placed in corn strip-cropped with alfalfa (▲), intercropped with fava bean (*), intercropped with apples (▼) and grown in monoculture (■) in Tlaxcala; B. Numbers of lycosid spiders caught in pitfall traps placed in corn rows bordering alfalfa strips and in rows located in the center of the field.

is subjected in Tlaxcala. Although the farmers may not be aware, we found that the production of corn in polycultures and agroforestry patterns, triggers a series of ecological interactions with important consequences for insect pest management and soil fertility relations.

Our studies showed that trees integrated into cropping systems modified the growing environment of associated understory corn plants, influencing their growth and yields. The greatest modification results from the interception of solar radiation, redistribution of precipitation and reduction of wind speed by tree canopies and deposition of leaf litter. Below ground, extensive root systems help to stabilize the soil, draw nutrients to be deposited later on the soil surface with litterfall and contribute organic matter to the soil as a result of root decomposition.

With decreasing distance from trees, we observed an increase in soil organic matter, total nitrogen, available phosphorus, exchangeable potassium, calcium and magnesium, cation exchange capacity and water holding capacity.

Certain tree species affected maize yields more than others. Such trees as apple, pear-apple and peach, with smaller crowns and lower stature, appeared to have a minimal influence on yields, thus acceptable harvests were obtained from corn intercropped in fruit orchards. Although corn monoculture systems in the region may exhibit significantly greater yields per hectare than most traditional agroforestry systems, when other considerations such as the riskiness, multiple protective and productive the functions, and returns from the most limiting input (e.g., labor, capital) are examined, the yield superiority of the monoculture becomes less clear-cut and attractive. Further studies can provide important information for determining which trees to encourage in fields and in what spatial designs, in order to enhance the multiple use capacity of the system.

Densities of the pest _Macrodactylus_ and of foliage and soil predators fluctuated depending on the arrangements of crops in time and space, the composition and abundance of non-crop vegetation within and around fields, the surrounding environments and the type and intensity of management. Thus _Macrodactylus_ sp., predaceous beetles, lycosid spiders and _Orius_ sp. responded depending on their degree of association with one or more of the vegetational components in the systems. For example, the abundance of insect predators and spiders in corn depended greatly on the presence and phenology of adjacent alfalfa strips. A greater abundance of predators in corn rows adjacent to alfalfa strips corresponded with a lower incidence of _Macrodactylus_ in those rows. Since alfalfa serves as a major source of natural enemies, correct designs of corn-alfalfa strip cropping systems and proper timing of alfalfa cuttings could greatly increase the abundance and efficiency of beneficial arthropods in corn systems. Also the corn-fava bean-squash polyculture exhibits inherent elements of crop protection that warrant further research to determine optimal planting times, planting densities and spatial designs of the component crops for improved biological control. In addition, infestations of this pest on corn can be ameliorated by encouraging other host plants within or around the fields (i.e., _Brassica_ spp., _Lupinus_ spp.) that serve as trap crops, enticing _Macrodactylus_ away from corn, as traditionally done by local farmers.

7.4. COMPARATIVE INSECT ECOLOGY OF APPLE ORCHARDS IN NORTHERN CALIFORNIA

During 1982-1983 we conducted studies to compare the species diversity, abundance patterns, and damage levels of insect pests and associated natural enemies in four different, dry farmed, California apple orchards: an 'abandoned' orchard not managed or disturbed for 25 years, two 'organic' (not sprayed with synthetic pesticides) orchards, one kept clean cultivated, the other with a cover crop, and a 'commercially' managed orchard (clean cultivated and managed with fertilizers and pesticides). Obviously these orchards are ecologically very different and constitute what may be termed a 'cultural evolution continuum'. The abandoned orchard may be considered as a substrate where the lack of disturbance has probably allowed the consolidation of consistent relationships between the resident insect fauna and the local vegetation. This agroecosystem exhibits characteristics of both agricultural and natural systems (i.e., non-managed, evenly spaced apple trees intermingled with plants typical of secondary successional stages), thus, presumably, many of the imbalances typical of a commercial apple monoculture system are significantly ameliorated in these orchards. The commercial orchard is a product of anthropocentric manipulations that substitute high energy inputs for some of the plant-insect interactions. The organic orchards combine characteristics of both systems.

The apple orchards are distributed within a matrix of natural vegetation which provided a unique opportunity for the study of arthropod colonization and exchanges of arthropods at the orchard-woodland interfaces. Research has shown that uncultivated habitats adjacent to agricultural areas can act as potential sources of colonization for pestiferous and beneficial arthropods (2). Many insect pests such as aphids and leafhoppers, invade crop fields and orchards from edge vegetation, especially when the wild plants are botanically related to the crop (22). On the other hand, the proximity of forest edges, hedgerows and patches of wild vegetation can influence the colonization, abundance and diversity of predators and parasites within a given agroecosystem, particularly when such ecotones provide hybernation and/or alternate feeding sites to natural enemies (37).

Despite this research, we still know little about the dynamics of arthropod populations at the interface of agricultural and natural plant communities. With our studies we attempted to answer some basic questions: a) to what extent do agricultural arthropods depend on non-crop habitats? b) do arthropods move from wild vegetation borders to nearby crop fields, and if so what triggers these movements? c) do these borders influence the species diversity and abundance of beneficial arthropods in adjacent crop fields? and d) will these colonizing arthropods remain within adjacent crops regardless of the degree of vegetational diversity and management intensity of these fields?

7.4.1. <u>The Effects of Orchard Structure</u>

Our results suggest so far that the incidence of codling moth, aphids and leafhoppers may be enhanced or reduced depending upon the type of orchard management, location, varietal diversity and the degree of vegetational complexity both within and in the periphery of the orchard. Pest populations and associated natural enemies seem to respond to plant successional features of the orchard depending on their degree of association with the cover crop, and surrounding vegetation. This is

certainly the case with Cicadellid adult and nymph populations, which
exhibited increasing densities with degree of orchard perturbation and
simplification (Figure 5). Simplified, insecticide treated orchards
exhibited highest leafhopper loads, whereas abandoned, vegetationally
diverse orchards were characterized by very low and stable populations. A
declining gradient in predator abundance on the apple trees was observed
from abandoned to sprayed orchards, paralleling a decrease in vegetational
diversity and an increase in management intensity.

Manipulation of ground cover vegetation in apple orchards and
vineyards had a substantial impact on the abundance of soil dwelling and
foliage inhabiting arthropods. Apple orchards with cover crops were
generally characterized by: (1) less colonization and subsequent
infestation levels of aphids, leafhoppers, and codling moths, (2) more
species and more individuals of soil dwelling predaceous arthropods, and
(3) higher removal rates of artificially placed prey. In contrast, disked
systems were generally characterized by greater herbivore loads on the
trees and by relatively low population levels of natural enemies per
habitat space (Figure 6a and b).

The cover crops harbored large numbers of prey (aphids, leafhoppers,
etc.) which attracted varying numbers of predators. However, high numbers

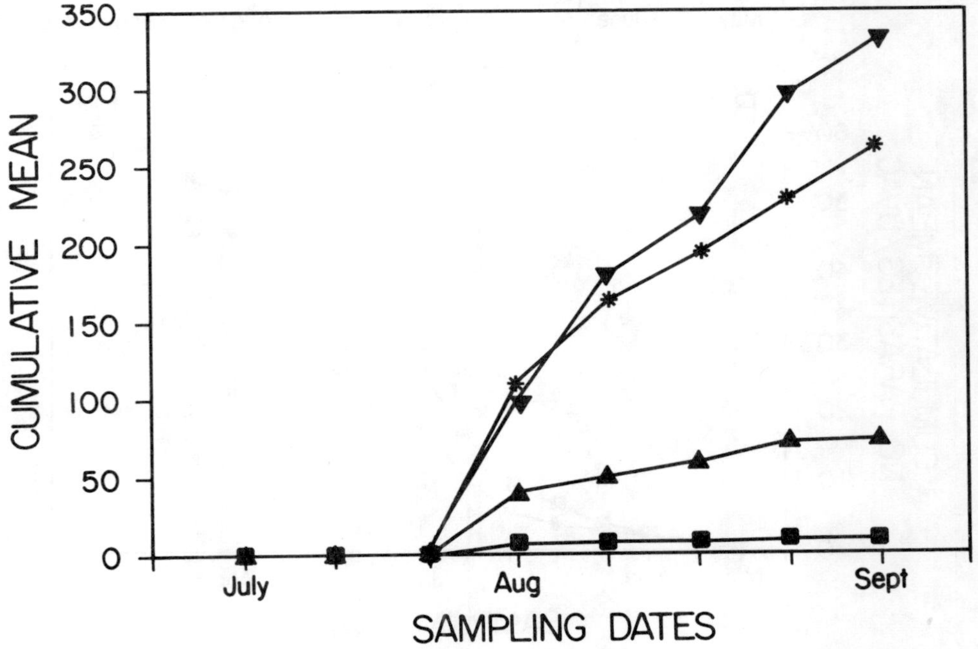

Figure 5. Cumulative mean densities of leaf hoppers (Homoptera:
Cicadellidae) on apple trees located within abandoned (■), organic with
cover crop (▲), disked organic (✳), and sprayed (▼) orchards in northern
California.

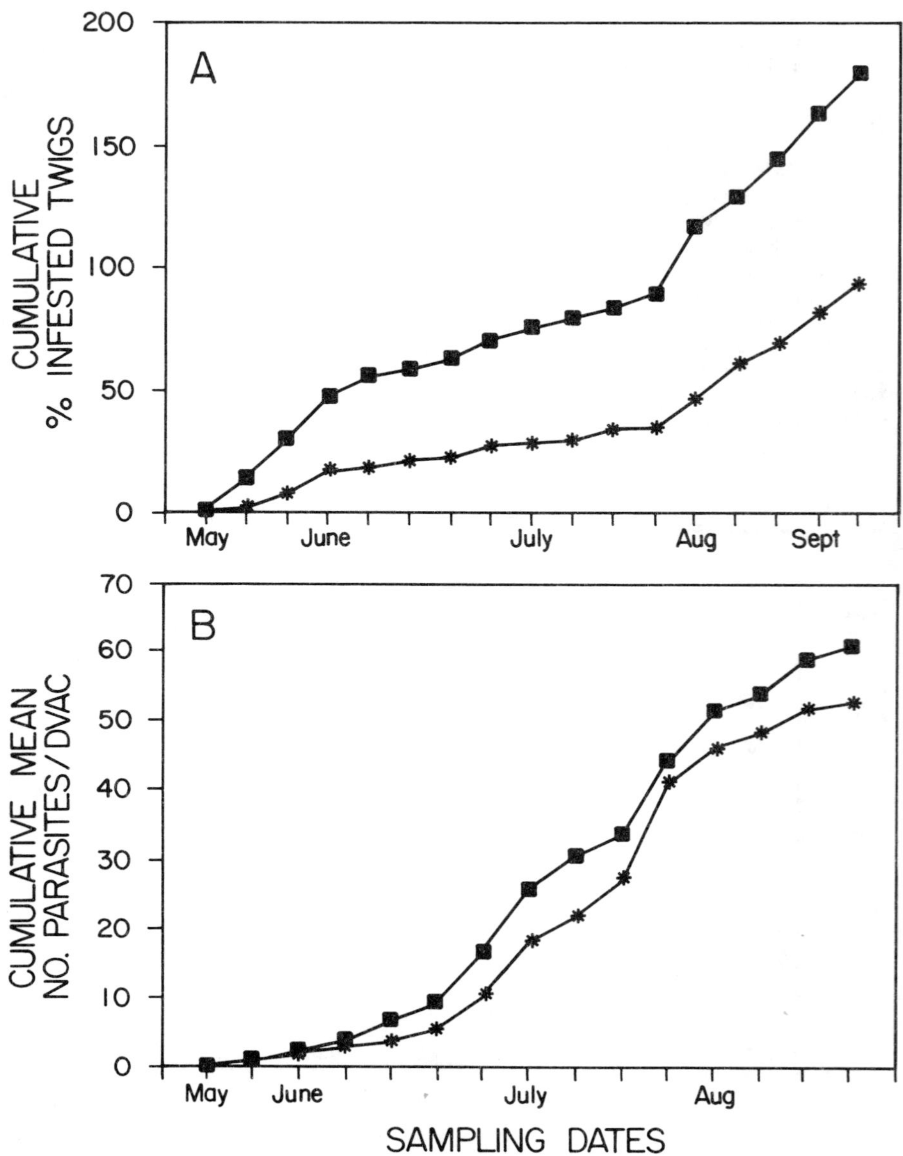

Figure 6. A. Cumulative proportion of twigs infested with leafhoppers in an organic apple orchard with (■) and without (✳) cover crops in northern California. B. Cumulative mean densities of parasitic Hymenoptera on apple trees in organic orchards with (■) and without (✳) cover crops in northern California.

of predators on the cover crops did not necessarily translate into higher numbers of predators on the trees, although cover cropping affected the dynamics of pest species on the trees. The data also do not indicate how realistically predation on artificial baits relates to reduction of actual apple pests, such as codling moths, aphids and leafhoppers.

The potential value of a cover crop as a reservoir of natural enemies seemed to be determined by the species of cover crop and its structural and phenological features. Legume cover crops that remained in full bloom throughout the season sustained the highest populations of arthropods. As the foliage of the various cover crops senesced and deteriorated nutritionally, supporting few phytophagous insects, the complex of predators and parasites narrowed in abundance and species composition (6).

7.4.2. Arthropod Dynamics at the Orchard-Woodland Interfaces

Considerable numbers of arthropods were intercepted by Malaise traps at the interface of abandoned, organic and sprayed apple orchards with adjacent woodlands (8). Higher catches of adult parasitic Hymenoptera and predaceous Raphididae, Hemerobiidae, Coccinellidae and Cantharidae were obtained at the interface of the disked organic orchard than in the sprayed orchard. Only Syrphidae were more frequently intercepted at the sprayed orchard than at the organic orchard- woodland interface. Low numbers of arthropods were collected in traps placed between the abandoned orchard and woodlands. Early season catches of immature and winged rosy apple aphids were considerably higher at the interface of the sprayed orchard than at the interface of the organic and abandoned orchards. Between mid-June and mid-July considerably more adult Coccinellidae were intercepted by the Malaise traps in the organic orchard than in the sprayed and abandoned orchards (Figure 7a and b).

Predators were more abundant on trees close to the woodlands than in the centers of the managed orchards. Removal rates by predators of artificially placed Anagasta kuehniella eggs, were higher on the trees in the field margins than on the trees in the centers of the organic and sprayed orchards. No differences in predation were observed between edge and center in the cover cropped and abandoned system (Table 2).

Herbivores exhibited an opposite trend. Apple trees located in the center of the organic orchards exhibited higher infestation levels of rosy apple aphids than trees located adjacent to the woodlands (Table 2).

7.4.3. Design Implications

Depending on the orchard system, cover crop complex, and associated arthropod species, it seems that manipulation of the ground cover can have a significant impact on the number of arthropods that inhabit the orchard ecosystem by: (1) directly affecting herbivore populations which discriminate among trees, with and without cover beneath, or (2) by attracting and retaining soil and foliage inhabiting natural enemies through the provision of alternate food and habitat. Encouraging cover crops that remain in full bloom throughout the season, that produce more biomass and support higher numbers of alternate host/prey, may be more appropriate to enhance the largest complex of predators and parasites. Experiments designed to test whether mowing the cover crop at different times forces movement of natural enemies up to the trees, will refine our

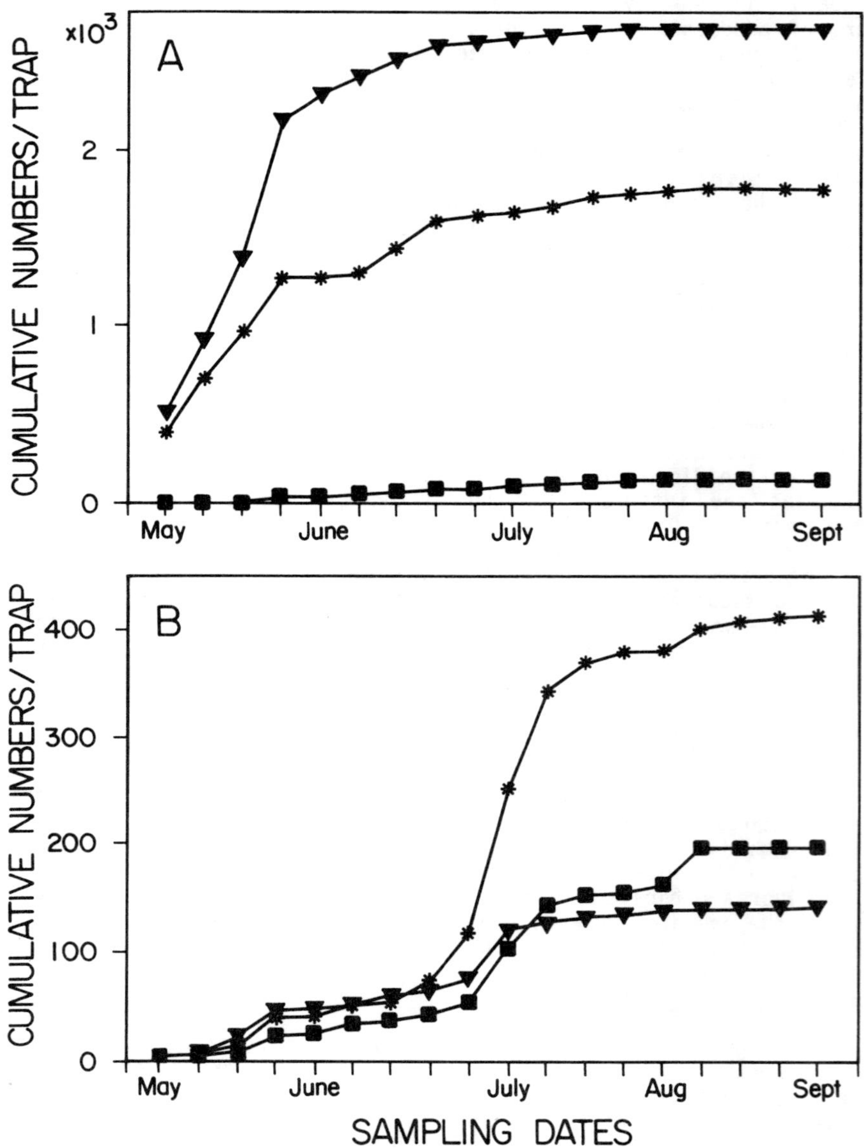

Figure 7. A. Cumulative number of rosy apple aphids (<u>Dysaphis plantaginea</u>) caught in Malaise traps placed at the interface of abandoned (■), disked organic (✻) and sprayed (▼) apple orchards and adjacent woodlands. B. Cumulative number of predaceous Coccinellidae (Coleoptera) caught in Malaise traps placed at the interface of abandoned (■), disked organic (✻) and sprayed (▼) apple orchards and adjacent woodlands in northern California.

104

Table 2. The effect of adjacent woodlands on the infestation levels of rosy apple aphids (_Dysaphis plantaginea_), the seasonal abundances of insect predators and predation pressure, in a range of apple orchards in northern California.

Orchard system	Apple trees adjacent to woodland			Apple trees in the center of the orchards		
	a	b	c	a	b	c
Abandoned	4.2±0.7	9.1±2.7	42.0±10.7	5.5±2.3	1.4±0.7	38.0±9.2
Organic cover cropped	6.8±2.1	6.0±1.6	36.9±11.7	17.1±6.7	1.4±0.8	29.3±8.0
Organic disked	9.8±3.2	9.5±3.4	34.1±9.1	19.4±5.9	1.6±1.1	25.0±6.2
Sprayed	20.2±5.6	2.3±1.3	26.0±8.0	26.9±8.9	0.6±0.3	21.0±3.2

a = % aphid infested twigs. Seasonal means ± S.E. derived from five sampling dates in May-June 1982. Twenty twigs per apple tree (5 center and 5 edge trees per orchard) were sampled on each occasion.

b = Adult Coccinellidae, Cantharidae, Chrysopidae, Hemerobiidae and predaceous Hemiptera. Seasonal means ± S.E. derived from 18 sampling dates in 1982. Five trees in the center and border of each orchard were sampled with a D-Vac on each occasion.

c = % eggs removed. Predation pressue on the eggs was estimated on four occasions by hanging twenty-five 8.5 x 11.0 cm paper cards (with 50 moth eggs each) for 18 hours from the branches of each of five trees in the center, border and edge or each orchard.

knowledge about cover crop manipulation as an insect habitat management technique in agroforestry systems. Critical testing of these effects in a range of orchard and vineyard systems is of substantial theoretical and practical interest. Biological control of certain orchard pests could be significantly enhanced through knowledge of these interactions.

Results from our studies on arthropod dynamics at the interface of the various orchards and the adjacent woodlands can be of great use for designing windbreaks or shelterbelts to surround orchards. Since several arthropod species collected within the orchards, it seems that the development of apple orchard arthropod communities is influenced by surrounding natural vegetation. A more precise knowledge of these interactions can lead to management strategies based on: a) manipulation of plant composition and abundance of field margins, b) change of the temporal permanence of these margins and therefore the time available for colonization, and c) modification of the distance of trees to the borders.

Advanced succession in the abandoned orchard resulted in increased natural enemy abundance, possibly a consequence of increased diversity of resources and alternate food. Herbivore densities were reduced, but it is not clear whether this is due to the increased abundance of natural enemies or to changes in tree growth (lower foliage levels of nutrients, scarce new tissue growth, etc.) which in turn may have affected herbivore reproduction. Based on these results, we feel that measuring component interactions in abandoned orchards provides important ecological information that may be applicable to modern fruit-tree based agroforestry systems. Croft and Hoyt (21) caution, however, that studying species only present in undisturbed systems, and that do not occur in commercial orchards, provide only limited information applicable to modern orchards. Further MacLellan (31) states that because commercial growers need to know the extent of upward levels of fluctuations of pest species if insecticides are not used, a study of the arthropod species in abandoned orchards, in which fruit is useless for trade purposes, fails to reflect population changes in commercial orchards with limited insecticidal use. Nevertheless, the study of K-type strategist species (i.e., apple maggot, codling moth) that are subject to the same regulatory processes in both unmanaged and in intensively managed orchards, might prove useful in identifying potential natural controls.

7.5. CONCLUSIONS

The central issue in sustainable agriculture is not achieving maximum yields, it is long-term stabilization. Increasingly, research evidence suggests that yields are more stable in agroecosystems where crops are grown in diversified patterns, or where several varieties of the same crop are combined in a field (33). Since Tlaxcalan traditional corn agroecosystems and abandoned apple orchards in northern California exhibit some ecological properties that result in damped oscillations in pest populations, we feel compelled to further examine these systems as starting points in our search for principles governing sustainability in agriculture. Apparently in our systems, vegetational diversity is a good strategy to cope with seasonality and variability. In the traditional system a wide ecological tolerance is attained in cropping systems by farmers that rely on a large number of agroforestry management strategies. In the abandoned orchards, natural ecological succession restores environmental balance and productivity.

The challenge is to determine whether low-input agricultural systems are, across the board, good models which include properties of constancy of production, self-regulating mechanisms and sustained soil fertility. Concurrently, in developing countries, the elements of traditional agriculture that should be retained in the course of agricultural modernization must be identified.

ACKNOWLEDGEMENTS

Studies mentioned in this paper were partially funded by grants of the University of California Consortium on Mexico and the United States and the Jessie Smith Noyes Foundation in New York. We thank John Farrell for allowing us to use some of his data from his Tlaxcala studies, ad also Linda L. Schmidt for elaborating the graphs and Joanne Fox for typing the manuscript.

REFERENCES

1. Alcorn, J.B. 1984. Huastec Mayan ethnobotany. Univ of Texas Press, Austin, TX.
2. Altieri, M.A. and D.K. Letourneau. 1982. Vegetation management and biological control in agroecosystems. Crop Protection 1:405-430.
3. Altieri, M.A. 1983. Agroecology: the scientific basis of alternative agriculture. Div Biol Control, Univ California, Berkeley, CA.
4. Altieri, M.A., Letourneau, D.K. and J.R. Davis. 1983. Developing sustainable agroecosystems. BioSci 33:45-49.
5. Altieri, M.A. and J. Farrell. 1984. Traditional farming systems of south- central Chile, with special emphasis on agroforestry. Agrofor Syst 2:3-18.
6. Altieri, M.A. and L.L. Schmidt. 1985. Cover crop manipulation in northern California orchards and vineyards: effects of arthropod communities. Biol Agric Hortic 3:1-24.
7. Altieri, M.A. and M.K. Anderson. 1986. An ecological basis for the development of alternative agricultural systems for small farmers in the Third World. Amer J Altern Agric 1:30-38.
8. Altieri, M.A. and L.L. Schmidt. 1986. The dynamics of colonizing arthropod communities at the interface of abandoned, organic and commercial apple orchards and adjacent woodland habitats. Agric Ecos Envir 16:29-43.
9. Altieri, M.A., Trujillo, J. and J. Farrell. 1986. The agroecology of corn production in Tlaxcala, Mexico. Human Ecol (submitted).
10. Alverson, H. 1984. The wisdom of tradition in the development of dry-land farming: Botswana. Human Organiz 43:1-8.
11. Bayliss-Smith, T.P. 1982. The ecology of agricultural systems. Cambridge Univ Press, Cambridge, U.K.
12. Beets, W.C. 1982. Multiple cropping and tropical farming systems. Westview Press, Boulder, CO. 156 p.
13. Benitrez, F. 1953. Los Primeros Mexicanos. Ediciones Era, Mexico. 536 pp.
14. Brady, N.C. 1984. The nature and property of soils. MacMillan Pub Co, New York. 750 p.
15. Brokenshaw, D.W., Warren, D.M. and O. Werner. 1980. Indigenous knowledge systems and development. Univ Press Amer, Lanham, MD.
16. Chambers, R. 1983. Rural development: putting the last first. Longman, London. 246 p.
17. Chang, J.H. 1977. Tropical agriculture: crop diversity and crop yields. Econ Geogr 53:241-254.
18. Clawson, D.L. 1985. Harvest security and intraspecific diversity in traditional tropical agriculture. Econ Bot 39:56-67.
19. Conway, G.R. 1985. Agroecosystem analysis. Agric Admin 20:31-55.
20. Cox G. W., and M.D. Atkins. 1979. Agricultural ecology. W. H. Freeman and Co, San Francisco, CA. 214 p.
21. Croft, B.A. and S.C. Hoyt. 1983. Integrated management of insect pests of pome and stone fruits. John Wiley and Sons, New York, NY.
22. Dambach, C.A. 1948. Ecology of crop field borders. Ohio State Univ Press, Columbus, OH. 203 p.
23. Edens, T.C. and D.L. Haynes. 1982. Closed system agriculture: resource constraints, management options and design alternatives. Ann Rev Phytopath 20:363-395.
24. Ellen, R. 1982. Environment, subsistence and system. Cambridge Univ Press, Cambridge. 324 p.

25.   Ewel, J.S., Gliessman, S.R., Amador, M. A., Benedict, F., Berish, C.,
      Bermudez, R., Brown, B., Martinez, A., Miranda, R. and N. Price.
      1984. Tropical agroecosystem structure. AgroEcos 9:185-190.
26.   Farrell, J. 1984. The role of trees in Tlaxcala farming systems. MS
      Thesis, Univ California, Berkeley, CA.
27.   Gliessman, S.R., Garcia, R. and M.A. Amador. 1981. The ecological
      basis for the application of traditional agricultural technology in
      the management of tropical agro-ecosystems. AgroEcos 7:173-185.
28.   Hernandez, X.E. 1977. Agroecosistemas de Mexico. Colegio de
      Postgraduados, Chapingo. 475 p.
29.   Klee, G.A. 1980. World systems of traditional resource management.
      John Wiley and Sons, New York, NY.
30.   Lawrence, R.B., Stinner, R. and G.J. House 1984. Agricultural
      ecosystems. John Wiley and Sons, New York, NY. 233 p.
31.   MacLellan, C.R. 1977. Trends of codling moth (_Cydia pomonella_)
      populations over 12 years on two cultivars in an insecticide free
      orchard. Can Entomol 109:1555-1560.
32.   Marten, G.G. 1986. Traditional agriculture in southeast Asia: A human
      ecology perspective. Westview Press, Boulder, CO. 390 pp.
33.   Plucknett, D.L. and N.J.H. Smith. 1986. Sustaining agricultural
      yields. BioSci 36:40-45.
34.   Rambo, T.A. and P.E. Sajise. 1984. An introduction to human ecology
      research on agricultural systems in southeast Asia. East-West Envir
      Policy Inst, Honolulu, HI. 327 p.
35.   Rappaport, R.A. 1986. Pigs for the ancestors: Ritual in the ecology
      of a New Guinea people. Yale Univ Press, New Haven, CT.
36.   Rhoades, R.E. 1984. Breaking new ground: Agricultural anthropology.
      Inter Potato Center, Lima, Peru.
37.   van Emden, H.F. 1965. The role of uncultivated land in the biology of
      crop pests and beneficial insects. Sci Hort 17:121-136.
38.   Wilken, G.C. 1977. Integrating forest and small-scale farm systems in
      middle America. AgroEcos 3:291-301.
39.   Williams, D.E. 1985. Tres Arvenses solanaceas comestibles y su
      proceso de domesticacion en Tlaxcala, Mexico. MS thesis, Colegio de
      Postgraduades, Chapingo, Mexico.
40.   Zinke, P.J. 1962. The patterns of influence of individual forest
      trees on soil properties. Ecol 43:130-135.

8. AGROFORESTRY PRACTICES AND RESEARCH IN INDIA

K. G. Tejwani

Resident Associate, International Center for Integrated Mountain Development, P. O. Box 3226, Kathmandu, Nepal

8.1. ABSTRACT

Many agroforestry practices in India are old and traditional. Some systems have been researched since the late 19th century, and there has been research on fodder-fuel plantations since the 1950's. A wide variety of practices in India are described here with research results incorporated wherever available. It is concluded that: (1) there are many systems and practices that are yet to be adequately described; (2) even though some systems are well-described, there is need of greater research on practices; (3) the research endeavor so far, apart from indicating possibilities for increased production in fodder-fuel plantations has not really improved the components of agroforestry systems and their management. The challenges and opportunities for improvement are vast.

8.2. INTRODUCTION

Many agroforestry systems in India are very old and traditional (e.g., shifting cultivation, taungya, growing tea and coffee under shade trees, intercropping under coconut, etc.). The only thing new is the use of the term "agroforestry." Not only have these systems been widely practiced, but some have also been widely researched (e.g., growing tea under shade trees since the latter part of 19th century; fodder-fuel plantations since the 1950's).

In India, the major land uses are classified as cultivated land, forest land, pasture land, and waste land. "Waste land" primarily refers to land unfit to grow agricultural crops, even though the land may be suitable for growing grasses and trees, and it could be termed "wasted land" instead.

Out of a reporting area of 306 million ha in India in 1982 (of a total geographic area of 329 m ha), there were 67.4 million ha of forest land, 16.9 million ha of culturable waste lands and 12.1 million ha of permanent pastures and other grazing lands (43). Forest land is mostly owned and managed by the State Governments, though some land is owned and managed by villages. Each village or group of villages is allotted some land. Most of the forest land is controlled by the forest departments of State Governments. People have rights in these forest lands to graze their animals and collect fodder, fuel and timber as per prescribed rules and regulations. In all the cases discussed, the trees in the system generally belong to whomever owns the land.

For more than three decades, India has made massive investments in agriculture and has achieved spectacular success in food production. Currently, its attention is becoming more focused on the problems of acute shortages of fodder, fuel wood and other forest products. Not only do the current land use systems (agriculture, forestry, animal husbandry) need more intensive research and development effort, but other traditional systems and new variations need to be explored. This chapter catalogues the agroforestry practices in vogue, describes the research efforts made so far, and points to new directions for research and development of agroforestry practices in India.

8.3. DEFINITION AND CLASSIFICATION OF AGROFORESTRY SYSTEMS

Combe (15) has suggested 24 classes of agroforestry systems, based on the kinds of associated agricultural products, major functions of the tree component, spatial arrangement of trees and duration of the combination. Nair (50), on the other hand, has suggested including agroecological zones as well as socio-economic aspects (e.g., silvopastoral systems for cattle production in tropical savanna or agrosilvicultural systems for soil conservation in tropical savanna). This paper will follow Combe's classification, since Nair's classification is an elaboration of it. Combe's classification is still incomplete, as apiculture and pisciculture with trees are more difficult to classify in his scheme. In case the trees are the major component of land use and an agricultural crop is integrated with them, the system is referred to as "Silvoagriculture" in this chapter. In case the agriculture crop is the major component and trees are secondary, the system is called "Agrosilviculture". A combination of trees and grasses is referred to as "Silvopastoral" if the grazing land is in forest, and "Pastoralsilviculture if trees are scattered in a pasture/grass land. If we consider that trees also grow in and around wetlands and other bodies of water, in which fish, etc., can be grown, the combination can be called "aquasilviculture." If we consider honey, flowers, oil, fruits, food, medicinal herbs/plants, lac, silk, etc., collected from trees as agricultural produce, one could classify all these practices as "Silvoagriculture" systems. This procedure expands Combe's (15) system to 48 classes of agroforestry.

8.4. AGROFORESTRY PRACTICES AND RESEARCH IN INDIA

Most of the practices described in this paper are indigenous and have been developed and practiced by the farmers in various agro-ecological regions, although organized research in agroforestry, primarily on growing shade trees and tea together, is over 100 years old. There has also been extensive research on intercropping under coconut, and growing of coffee and rubber under shade trees. Some experiments on pastoralsilviculture systems (referred to as fodder-fuel systems in India) were started in the 1950's. A large number of institutes have been conducting research on agroforestry systems without using the term "agroforestry". Realizing its importance, an all India Coordinated Research Project on Agroforestry was initiated in 1983 to initially operate at eight Research Institutes of the Indian Council of Agricultural Research (ICAR) and 12 Agricultural Universities. This project was considerably expanded in the Seventh Five Year Plan (1985-90).

8.4.1. Silvoagriculture Systems

Shifting cultivation

This system is practiced extensively in the north eastern hill region (comprising the states of Assam, Meghalaya, Manipur, Nagaland and Tripura and two union territories of Arunachal Pradesh and Mizoram) and to some extent in Andhra Pradesh (AP), Bihar, Madhya Pradesh (MP), Orissa, and Karnataka states (Fig. 1). It is called "Jhum" in north east hill region and "Podu" in AP and Orissa States. Considering the steep landscape, high rainfall, and isolation of the hill tribes, shifting cultivation is the most appropriate response to the agroecological situation. The entire way of life, training of youth, ceremonies and festivals, and the philosophy are a product of this system (75). The climate is tropical to sub-tropical with high rainfall and high humidity (rarely below 75% throughout the year). Forest types are tropical evergreen, temperate evergreen, conifer, and pine forests. Estimates prepared in 1974 indicated that about 2.7 million ha are used for shifting cultivation in the northeast hill region, of which 16.8% is cultivated at one point in time (6). The number of years before a community returns to the same land will depend on the land available to the community and its population. Currently the cycle has shrunk to 5-7 years as compared to earlier decades when it was as long as 30-40 years. With the present trend of steady rise in the population of the region (decennial population growth rate of 35% in 1971-81) (12, 54), the cycle is bound to shrink further, threatening economic as well as ecological aspects. Out of 4,343,000 tribals (in 1971), as many as half were engaged in shifting cultivation in the region (6).

The average size of plot cultivated by a family of 2 adults and 3 to 4 children is 1.0 to 2.5 ha. There are significant differences in the practices in different geographical areas. The seed mixture, as well as length of the cycle, vary considerably from region to region. Some 8-13 crops are grown in the same field. The longer the cycle (10 to 15 years), the greater the variety of crops grown; in a shorter cycle (5 years) emphasis is on maize, a leafy vegetable and tuber crop (39). The system is very labor intensive, requiring very little capital and inputs. The labor is provided by the whole family. The chores to be performed by the males and females are well-defined. Males contribute more to forest cutting and crop protection and females contribute more to burning, cleaning, sowing, weeding and harvesting. However, both participate in all the operations (64). The yields of crops decline with shortening of the cycle (39).

If the system is practiced over a cycle of 10-15 years, it has both protection and production benefits. Herbaceous species dominate during the first five years of fallow; these are then replaced by bamboo (Dendrocalamus hamiltonii), which progressively increases both in frequency and density in a 20 year fallow (62). In a 50-year fallow, bamboo gives way to broad-leaved tree species. Mixed cropping accompanied by successive maturing and harvesting of crops over a period of time has a good protective benefit during cropping. The system also provides built-in "insurance" to the farmer: one or the other crop(s) are likely to give good return under adverse climatic conditions. The early stage of shifting cultivation is a major cause of soil erosion; annual soil loss is reported to vary from 3.3 to 201 tonnes/ha (77).

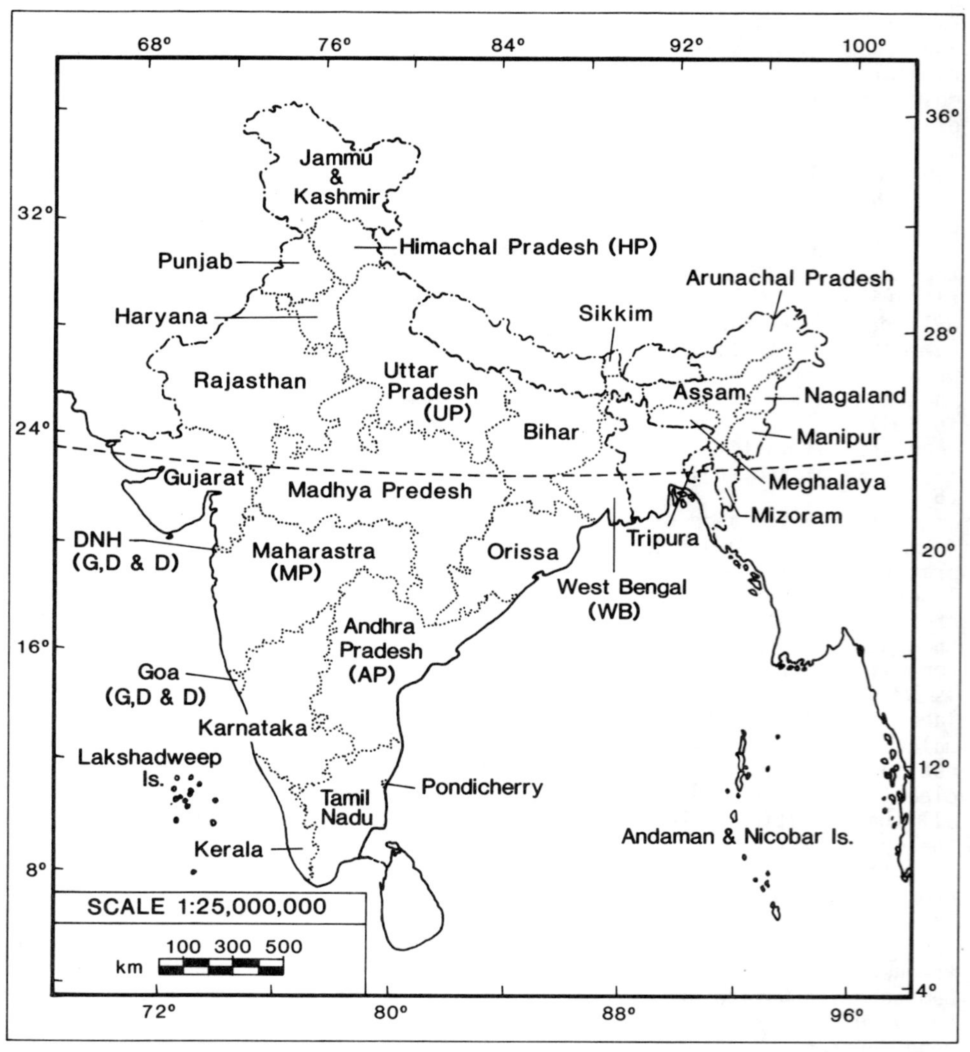

Fig. 1. Locations of the various states in India as referred to in this chapter.

The ICAR research complex for the North Eastern Hill Region has suggested a model land use plan for the North Eastern Hill Region as an alternative to shifting cultivation (76). This provides for the top 1/3 of the hill slope to be under forest, the middle 1/3 to be in horticulture, and the bottom 1/3 to be under bench-terraced cultivation (76). Forestry and horticulture provide protective cover to the steep land, and the lower 1/3 benefits from the run off, fertile soil, and plant nutrients received from upslope. This plan still needs wider demonstration and field trials for testing its adaptability, economic viability and social acceptability. The plan also requires a regular supply of seeds, fertilizers, and plant protection materials, and trained personnel for extension to a degree that may prove difficult to accomplish.

Taungya

Taungya is a system of establishing commercial forest plantations in which agricultural crops are grown on a temporary basis between regularly arranged rows of tree species. The system was introduced to India by Brandis, in 1956 and the first taungya plantations were raised in 1963 in north Bengal (69, see also Lowe, this volume). It is practiced in the states of Kerala, West Bengal (WB), and Uttar Pradesh (UP), to a little extent in Tamilnadu, AP, Orissa, Kranataka and north eastern hill region. In south India the system is called "Kumri." It is practiced in areas with assured annual rainfall of over 1200-1500 mm.

The practice consists of land preparation, tree planting, growing agricultural crops for 1-3 years, and then moving on to repeat the cycle in a different area. In some cases crops may be grown one year before the trees are planted. A large variety of crops and trees, depending upon the soil and climatic conditions, is grown (Table 1). The owner of the land, and hence the trees, is the State Forest Department.

Establishment of teak (_Tectona grandis_) plantations is the most renumerative taungya practice for the forest departments as well as the farmers. Due to prevailing land shortages as well as the possibility of growing highly remunerative crops (like paddy, tapioca, ginger, and turmeric), the farmers are willing to pay as much as Rs. 500- 1000/ha to lease of the land (47). In the high hills of Tamilnadu, _Eucalyptus globulus_, _Acacia mearnsii_ and pine plantations are established by growing an initial cash crop of potatoes. More recently, cashew plantations in the coastal belt of AP are established by cultivating groundnut. It is possible to grow two crops of groundnut in a year, and for conservation purposes, it is a condition of the lease that groundnut leaves are not removed from the land (63). Since 1962, large commercial farmers have used taungya for the establishment of poplars in the Terai region of UP with irrigation.

Resource input and utilization vary from state to state. As already mentioned, in some states there is willingness to pay large sums of lease money. Forest departments establish the tree plantations and the cultivators provide the labor to keep the plantations clean and free of weeds while cultivating the crops. Usually all the inputs for the crops are provided by the farmers. Farmers may also be required to provide labor for cutting the bridle paths, and in some cases the forest department may provide rudimentary facilities for the farmers to stay near the plantations. If the crop receives fertilizers the tree crop benefits considerably. Working the soils and weeding the agricultural crops also

Table 1. Crops grown in taungya system in India.

State	Tree crop	Associated agricultural crops
Uttar Pradesh (UP)	*Shorea robusta* *Tectona grandis* *Acacia catechu* *Dalbergia sisoo* *Eucalyptus* spp. *Populus* spp.	Maize, paddy, sorghum, pigeon pea, soybean, wheat, barley, chickpea, rape seed and miscellaneous
Andhra Pradesh (AP)	*Anacardium occidentale* (cashew) *Tectona grandis* *Bombax ceiba* Bamboo *Eucalyptus* spp.	Hill paddy, groundnut, sweet potato
Kerala	*Tectona grandis* *Bombex ceiba* *Eucalyptus* spp.	Paddy, tapioca, ginger, turmeric
Assam	*Shorea robusta*	Paddy
Tamil Nadu	*Tectona grandis* Bamboo *Santalum album* *Tamarindus indica* *Acacia nilotica* *Acacia mearnsil* *Ceiba pentandra* Cashew Rubber	Millets, pulses, groundnut, cotton, tapioca, potato
West Bengal (WB)	*Tectona grandis* *Shorea robusta* *Schima wallichii* *Cryptomeria japonica* *Quercus* spp. *Michelia doltsopa*	Paddy, maize, millets, turmeric, ginger, ladys finger, pineapple, hemp
Maharastra		Sunhemp, jute, mesta, sunflower, castor, etc.
Andaman and Nicobar Islands	*Pterocarpus dalbergoides*	Sugarcane, maize

benefit the trees. If agricultural cropping is prolonged, not only are the crop yields reduced, but in some cases it may also be harmful to the trees. For example, a definite loss of teak growth has been demonstrated in south India for each successive agricultural crop, even the first (47). In the case of cashew plantations, the groundnut leaves, which are left on the land, benefit the trees. Cashew itself is capable of binding sand, thus preventing its drift. Even though the farmers still cultivate the land up and down the hill slopes, causing severe erosion, it is felt that the damage done in 2-3 years of cropping is compensated for by the forest growth over the next 40-80 years.

Taungya has both positive and negative socio-economic aspects, depending upon the situation. Due to prevailing land shortages, the system provides an opportunity to a class of people to cultivate land and to obtain employment and food. In some cases, it is highly remunerative to the farmers. In all cases the system is highly remunerative to the forest departments, which are able to establish tree plantations with not much cost to themselves. However, the system is now considered to be highly exploitative of the poor people, who have to live the life of nomads at very low wages. Due to recent awareness, there is a tendency not to hand the land back to the forest departments. As a consequence, forest departments are also wary of practicing taungya. The system could still be viable and useful if a fair deal is provided to the farmers. There is also the potential for mechanization of the system, and comparisons of the mechanical and manual systems for very large scale afforestation; no research has been done on this topic.

Growing agricultural crops in combination with commercial trees

Commercial nut and fruit trees take a long time to mature and yield cash returns; hence growers look for ways to generate interim returns. These trees are usually planted in widely spaced rows. Until they grow and cover the entire land, it is possible to grow agricultural crops in the interspaces. As a consequence, growing agricultural crops in combination with commercial tree crops is a very widely practiced system. The most outstanding examples are intercropping with coconut (49), arecanut (52) and horticultural trees.

Intercropping with coconut: Coconut (Cocos nucifera) grows in eastern and southern India with temperature determining the boundaries with respect to altitude and latitude. Kerala, Tamilnadu, Karnataka and AP in south India account for over 90% of the area and total production of coconut. A mean annual temperature of 25°C with a diurnal variation of 6-7°C is optimal. It does well if the annual rainfall is 1500-2250 mm and is well-distributed (38).

Coconut is a perennial tree, which due to its height, crown shape, root pattern, and wide space of planting, is amenable to the growing of annual, biennial and perennial crops under it. The life span of the palm could be divided into three distinct phases from the point of view of intercropping: (i) within 8 years of planting, when the crown size increases gradually and the interspaces could be utilized for intercropping with annuals or short duration crops that do not compete with the developing palm; (ii) the period from 8 to 25 years, when the canopy covers 80% of the ground and is also low due to short trunks of the palm; there is little or no scope for growing other crops in the interspaces; and (iii) the period after 25 years, the most important from the point of view of intensive cropping. After 25 years, there is a gradual increase in the magnitude of light penetration to the ground due to increased trunk height and decreased ground coverage by the canopy. Under normal management conditions, the roots of an adult, bearing-palm, growing in a medium-textured soil (sandy loam) are concentrated within a radius of 2 m around the tree. Also, the top 30 cm of soil are practically devoid of functional roots (32). In a pure stand of coconut about 75% of the land area is not being utilized by the coconut roots (49).

A wide variety of crops is grown with coconut palms. Coconut is essentially a crop of small growers in India over 90% of the coconut holdings in south India are less than one ha and the average is only 0.22 ha. Hardly 2% of the holdings have an area of two or more ha. The labor required for the crop can be spread fairly evenly over the year. A great number of holdings are owned by part time farmers or people engaged in non-farm work who consider it safer to invest in coconut holdings, which provide tax free income, and engage labor to do the farm work. Coconut plantations are labor intensive and normally no fertilizer or manures are added. However, there is experimental evidence to suggest that it is desirable to fertilize coconut as well as intercrops for best results (27, 52, 87, 88).

Data on yield of coconut and other crops growing with it in farmers' fields are not available. The only information available is on experimental plantations (88). Coconut and cacao are compatible and contribute to increased yield of both the crops. All the intercropping systems experimentally tested are highly productive in terms of income and employment generation.

Only certain experimental innovations are discussed above. There is considerable scope to improve this system. Many important aspects such as selection of compatible crops, spacings, water management, fertilizer application, soil fertility, pest and disease management, long term production and protective benefits/interactions and socio-economic aspects need to be investigated.

Intercropping with arecanut: India is the largest arecanut (Areca catechu), or areca palm, producing country in the world (75% of the total), producing about 150,000 tonnes of nuts from about 183,400 ha of land (7). The interrelation of temperature, rainfall and humidity determines the boundaries of the arecanut area in altitude and latitude. For good growth a temperature range of 18-38°C without extreme variation, and high annual, well-distributed rainfall of 2,250 mm appear to be necessary. It can stand rainfall of up to 4,000 mm, provided soils are well-drained. The crop needs to be irrigated during the dry season. The palm needs protection from exposure to sun scorch; this is achieved by deflecting the north-south planting line toward the west, and, when young, a shade crop of banana is often used.

Due to its height, crown shape and wide spacing, areca is also amenable to growing annual, biennial and perennial crops under it. Banana, cacao, black pepper, pineapple, betel vine, elephant foot yam, tapioca, turmeric, ginger and Guinea grass (Panicum maximum) are grown successfully under it.

In areca palm nurseries, banana is grown to provide shade to the seedlings. In case of intercropping with black pepper, areca palms of more than 10 years of age and 7-8 m tall serve as standards. In north

116

Bengal, farmers usually only take one crop. However, this seems mainly due to lack of technical know how, irrigation and availability of planting materials. The crops preferred as intercrops vary from tract to tract, though bananas are more universally grown (81).

For a steady and high yield of arecanut, growing in high rainfall areas with leached soils, annual split applications of inorganic fertilizers and green leaf and compost are recommended (except the green leaf and compost) (3, 7).

The yield of arecanut increases when intercropped with many crops (e.g., ginger, betel vine, black pepper, elephant foot yam, arrow root). The practice of intercropping is reported to be highly profitable although there is little information available on labor requirements.

Currently, due to changes for the better in socio-economic conditions such as the general level of farming, marketing, credit facilities, roads, and cooperatives in south India, the emphasis is on intensive land use and maximum productivity. In north Bengal the efficiency of production and marketing is still low. Even though intercropping under arecanut is still profitable, less than 15% of the area is intercropped (81).

This system has great potential with respect to increasing production. There is little information, however, about long term management. Information, planting materials, inputs and cash resources for farmers are still limited.

<u>Intercropping with horticulture trees</u>: It is estimated that there were about 2.1 and 2.5 million ha under fruit trees in India in 1976 and 1982, respectively. This represents a 19% increase in area in a period of six years and indicates that there are many young orchards. Mango is the most dominant fruit, occupying over 40% of the total area with apple, banana, guava, pineapple and other fruits 22.0% of the total area (44).

Some fruit plants are cultivated like field crops (e.g., banana, pineapple, papaya and strawberries) and are grown as intercrops among trees. Trees that take a longer time to bear fruit and are spaced widely, are amenable to intercropping, examples are mango, apple, citrus, litchi, date palm, almond, apricot, peach, plum and nectarine. Most of the orchards are irrigated and, although information is incomplete, species may require very different management practices (18, 82).

Though intercropping is generally practiced for economic reasons, neither detailed descriptions nor research results are available.

Rainy season vegetables, winter vegetables (with irrigation), grain beans, berseem (<u>Alexandrinum trifolium</u>), and other crops which do not grow too tall or require too much irrigation, are grown with mango, chiku, loquat, and pomegranate. Since only 60-100 mango trees are grown per ha, filler trees (e.g., peach, plum, guava) are also being grown. These filler trees are uprooted after 10 years. Papaya, phalsa, pineapple and dwarf varieties of banana can also be grown as intercrops depending upon the edaphic conditions.

Citrus can be intercropped with winter vegetables and berseem (with irrigation) and gram for 2-3 years. Apple varieties with crab seedling root stock (250 trees/ha) can be intercropped for 5-6 years with beans and peas. Apple varieties on semi-dwarf root stock, planted closely (2500 trees/ha), commence production in about 5 years; hence they are not amenable to intercropping. Apricot, peach, plum, nectarine (250 trees/ha) can be intercropped with beans and peas for 2-3 years. Cultural requirements of litchi are similar to those of mango, except that it is a shallow rooted tree; deep tillage or plowing is therefore harmful and

leguminous crops are preferred. Young orchards of litchi could also be
planted with filler trees like papaya and phalsa. Intercropping in date
palm plantations is possible with sufficient availability of water.
Saffron is cultivated in young almond plantations in Jammu and Kashmir
State, and bee keeping is practiced in orchards in Jammu and Kashmir, HP
and UP states.

Other specific agricultural practices associated with forests

If the term agriculture is used in its broadest sense, not only
covering crops, animal husbandry and horticulture but also pisciculture,
apiculture, floriculture, sericulture, lac culture, etc., a number of
"silvoagriculture" practices can be identified. These would include
systems of harvesting/collecting food, oil producing trees, dye producing
flowers, seeds and flowers (orchids), industrial raw materials (such as
gums, tannin, resin, soap nut, bamboo, canes, etc.), rearing of silk moth
and lac insects, and fisheries in association with mangroves, ponds and
tanks where trees also grow (51).
None of these systems has been studied and described. There is a
vast variety of trees, bushes and herbs in the forests used by people for
many products. These products, in the language of foresters in India, are
unfortunately referred to as "minor forest products". However, if one
were to take into account the value of these "minor forest products", and
if one were to take into account the vast numbers of humans which depend
on them, we might refrain from the use of this term.

8.4.2. <u>Agrosilviculture System</u>

Trees in agricultural fields

Trees are grown in agricultural fields for many reasons. Some of the
systems are very extensive and perhaps all of them are well understood by
the farmers who have developed them and have been practicing them
traditionally. Only a few important ones are described here.

<u>Prosopis cineraria</u> and <u>Zizyphus nummularia</u>: This system is extensively
practiced in the arid parts of Rajasthan and Gujarat states and to some
extent in parts of Punjab and Haryana bordering the Rajasthan desert. The
annual rainfall may vary from 200 to 500 mm. Once in 10 years the annual
rainfall of the arid zone may be less than 200 mm. Soils are low in
organic matter and have low water holding capacity. The system has been
described for Rajasthan by the Central Arid Zone Research Institute,
Jodhpur (4, 35, 74).
It is revealing that in the arid zone as much as 45% of the land is
cultivated (but only 1.45% double cropped) under rainfed conditions when
the land is classified as unsuited for agriculture. Surveys conducted
covering 64,140 km² (23.7%) of arid Rajasthan have shown that the present
land use is inconsistent with the land capability (34). Crop production
in this arid zone is unstable and risky, leading to low and unremunerative
yields. Cereals (<u>Pennisetum typhoides</u> and <u>Sorghum vulgare</u>), sesamum, and
beans (<u>Phaseolus radiatus</u>, <u>P. aconitifolius</u>, and <u>Cyamopsis tetrogonoloba</u>)
are the main rainfed crops. Livestock husbandry plays an important role
in the economy of the region, providing employment to two-thirds of the
population. Cattle, sheep, goats and camel are the most important
livestock species although their productivity generally is low (1). The

118

arid zone of Rajasthan has about 230,000 ha of tropical thorn forest in the rainfall zone of 250-700 mm. Thorny Mimosea occur throughout with a number of species, of which Prosopis cineraria is perhaps the most common; Acacias are widespread and Capparis decidua is often one of the most conspicuous trees. By arid zone standards, the Rajasthan desert is one of the most thickly populated deserts of the world, having an average density of 48 persons per km^2 as compared with an average density of 3 in most other desert regions of the world. The consequence of the increase in human and livestock populations, coupled with a lack of a diversified economic base, has been a tremendous pressure on land, water and plant resources (33). Size of individual land holdings varies from 3.27 to 24.6 ha, increasing with aridity (26). It is with this stark background that the practice of growing Prosopis cineraria trees in the agricultural fields is to be reviewed.

P. cineraria is a slow growing tree in the early stages. It takes 10-15 years to mature in the 200-350 mm rainfall zone, and 20 years in the 100-200 mm rainfall zone (35, 66). Slow rates of growth of the aerial part are one of the adaptations of this tree, as it initially establishes itself with a strong taproot system. The taproot penetrates the "kankar" (calcium carbonate) pan present at 50-150 cm soil depth (35, 66). The density of P. cineraria varies from 5 to 80 trees per ha depending upon the edaphic conditions of soil and rainfall; its density increases from western Rajasthan to eastern Rajasthan as the rainfall increases and soil conditions improve. The rainfall belt between 250-400 mm shows higher density (72). The plant has a high regeneration capacity.

Prosopis cineraria trees are planted on farm boundaries as well as in agricultural fields. Trees are lopped in a systematic manner from middle November to middle January every year (35, 66). In case the area is to be planted to wheat in winter, lopping is completed by the end of October or first week of November. The tender twigs are cut in April-June when fodder is scarce. The crowns of trees and agricultural crops grow simultaneously during the rainy season in July to September. The trees achieve full foliage by end of September, when the agricultural crops also start maturing. Although the trees are well-maintained by good farmers, in many instances the lopping is indiscriminate. The methods of propagation, growth and utilization of P. cineraria including lopping are based on traditional knowledge.

Every part of this tree is utilized by the people to meet their needs (45, 60). Fuel wood is of high calorific value and is used for making charcoal. Thorny twigs and branches are used for fencing the fields. The green and dry pods are used as vegetable, and the ripe pods are considered a fruit of the desert. One tree may yield as much as five kg of dry ripe pods. The leaves of P. cineraria provide nutritive green fodder; dry leaves are preserved and fed to livestock. Apart from dry leaves, small slender twigs with leaves are marketed in the form of small bundles. P. cineraria trees provide leaf fodder starting with 10th year and may go on producing up to 200 years. A moderately growing tree yields nearly 25-30 kg of dry leaves per annum. The flowers and bark are used as medicine, the bark as well as galls formed on the branches are used for tanning, and gum is used in sweets. Roots are used for making cot frames, handles for agricultural implements, rakes, bullock cart frames, butter churning sticks, etc. In view of this, the tree is called the "king of the desert."

The indigenous trees and shrubs in the arid zone of Rajasthan that have been identified as providing valuable forage on the basis of palatability ratings are Acacia nilotica; Prosopis cineraria; Zizyphus nummularia; Acacia senegal; Albizia lebbek; Anogoissus rotundifolia; Anogoissus pendula; Calligonum polygonoides; Azadirachta indica; Tecomella undulata; Grewia tanax; G. spinosa (19).

Studies have shown that the soil under P. cineraria has more organic matter, total nitrogen, extractable phosphorus, potassium, and micro-nutrients, and slightly lower pH and electrical conductivity than the soil under field conditions (2). Soil under P. cineraria also has larger populations of bacteria, fungi, actinomycetes and nitrifying bacteria than under Albizia lebbek, Tecomellia undulata and Prosopis juliflora. The observed improvement of plant growth under P. cineraria trees is thus due to the combined impact of better soil moisture and soil fertility. The additional possibility of contributions of bird droppings, leaf/pod fall, and dung and urine of shade-seeking livestock may also contribute.

This system also has many indirect protective, social and economic advantages. For example, P. cineraria, apart from improving soil fertility, acts as a soil binder, decreases the velocity of hot summer winds, provides shade to humans, livestock and birds during scorching heat in summer, and provides greenery in acute dry periods. It provides a source of income in drought periods and builds up a balanced economy and self-sufficiency of the farmers. The value of an agricultural field is determined by the number of trees grown, increasing with more trees.

P. cineraria is very much part of the cultural and socio-religious life of the people (at marriage, death, religious rites) (61, 71). The system may degrade if the trees are not properly lopped, but no case histories have been reported. The system could be improved by including fast-growing and high yielding P. cineraria trees. The scope of increasing production by joint management of the trees and crops has not been explored. Some of the research needs are for selection and evaluation of fast-growing trees with high foliage and food production commensurate with high protein content, high digestibility and low tannin content of leaves; improvement and development of already existing spineless trees of P. cineraria; development of agroforestry management systems with respect to the lopping cycle, lopping intensity, plant spacing, control of gall formation and insect pests, etc.; and economic evaluation of the system with respect to direct and indirect benefits (35).

Zizyphus nummularia: This is an important shrub which also grows and is maintained extensively in the cultivated fields in the arid zone. There is often a high density of this shrub (250-500 per ha) in cultivated fields with a well-defined kankar pan at 100-150 cm soil depth. Exposed gravelly plains support a density of 120-150 shrubs per ha (67).

In agricultural fields, the cutting of this shrub commences after harvest of kharif crops. Plants are cut at ground level and four to five plants are heaped together, and allowed to air dry. The air-dried foliage ("Pala") is removed by threshing. One coppiced shrub may produce 80-140 g of pala (67). Yearly cut shrubs on cultivated fields do not bear fruits. Zizyphus and agricultural crops can be grown together. In case of a drought year, a farmer is at least able to harvest Pala.

Grewia optiva and other species: This system is practiced in HP and UP Himalayan regions at elevations from 500 to 2300 m. The climate is sub-tropical to temperate, according to elevation, with distinct rainy

winter and summer seasons. Mean annual rainfall is 1000-2500 mm.
Topography and soils are variable. Major land use systems are
agriculture, horticulture, forestry and agroforestry. Agriculture is
subsistence with some cash crops like potato, ginger and beans.

Individual land holdings are usually less than 2 ha. These small
holdings are further fragmented and are at many locations; each fragment
could be further divided into very small fields (20, 70). Density of the
population is high, at 77 and 95 per sq. km in HP and UP, respectively in
1981 (12, 54).

Major agricultural crops grown are maize, paddy, wheat, barley and
potato; off season crops are tomatoes, cabbage and ginger. Main fruit
tree species are apple, plum, peach, apricot and mango (at lower
elevations). Fruit trees are also grown in agricultural fields in an
agro-horticulture system.

Trees grown by the farmers in the system for fodder, fibre, fuel and
timber are _Grewia optiva_ (300-2500 m), _Morus serrata_ (1000-2700 m), and
Celtis australis (500-2500 m).

Grewia optiva is a small to medium-sized deciduous tree; it
flourishes in regions of mean annual precipitation of 1200-2500 mm. It is
a multiple purpose tree and pollards and coppices well. Its bark yields
fibre, its leaves provide very nutritious fodder, the fruit is edible and
its wood is used for axe handles, shoulder poles, cot frames, etc. It is
seldom used as fuelwood because of its unpleasant odor (21). Leaves and
edible green twigs are palatable, nutritious and easily digestible.
Nutritious young leaves are converted into protein rich meal after drying
in sun. It is most often raised by planting seedlings at an 8 m spacing
in single rows on field bunds and terrace risers. It is very heavily
lopped, specially during winter, when no other green fodder is available
(61).

Celtis australis is a moderate to fairly large deciduous tree, which
can grow over a wide altitudinal range of 500-2500 m. Average annual
rainfall in its habitat varies from 1200-2500 mm. It pollards and
coppices well. It is raised primarily for fodder but is also used as
firewood, making tool handles and plows. Summer-lopped fodder has higher
nutritive value than autumn lopped fodder. The species is raised by
seedlings or rooted root-shoot cuttings (61).

Morus serrata is also a moderate to large deciduous tree, found in
mixed, lower western Himalaya forests at 1200-2700 m elevation. It is a
tree of sub-tropical to temperate climate; grows in a rainfall zone of 660
to 2970 mm, part of which is received as snow at temperate high
elevations. It is frost hardy, and is a good coppicer and produces root
suckers. It is raised for fodder, timber and fuelwood. Its leaves make a
valuable base for sericulture. Its wood is excellent for furniture
carving, cabinet work, sport goods, toys and agricultural implements. Its
fruit is sweet and edible. The tree is raised from seedlings, stumps or
rooted branch cuttings. It is lopped in summer (21, 61). _Moringa
oleifera_, _Quercus_ spp. and other minor species are also raised.

Grewia, _Celtis_ and _Morus_ are ready for lopping at the age of 10 years
and continue to be lopped for about 30 years. They are raised on field
bunds/terrace risers at 5-8 m linear spacing. The entire length of the
bunds/terrace risers is not planted systematically, with only about 1/3 of
the area planted. The average number of trees per farm holding is 30
(61). There is no specific mixture of trees and crops, and animals are
part of the system, including sheep, goat, buffalo, kine and some ponies.
Each family on an average, keeps 4 cattle. These domestic cattle are

partially stall-fed with grass (green or hay) and/or leaf fodder. For most of the time they are let out in the village grasslands and wastelands, adjoining forest areas, grass reserves (after harvesting grass) and fallow agricultural fields.

<u>Other tree spp.</u>: Acacias, <u>Eucalyptus</u> and many other species are also grown in agricultural fields, but detailed descriptions are not available. Since neither agricultural scientists nor foresters recognize these as components of crop or forestry systems, they remain neglected. Some useful information is emerging. Jambulingam and Fernandes (25) have described growing of <u>Borassus flabellifer</u>, <u>Tamarindus indica</u>, and <u>Ceiba pentandra</u>, with cereals and pulses by the farmers of Tamilnadu.
　　Farmers in temperate Nilgiri hills receiving high rainfall often grow <u>Eucalyptus globulus</u> on the outer edges or risers of bench terraces. The terrace crops are potato, cabbage and cauliflower. The trees are grown primarily for harvesting of leaves to extract oil. No economic analyses of this agroforestry system are available. It has been reported that the yield of potato is adversely affected by growing E. <u>globulus</u> (65, 68). Apparently the farmer knows better since he is willing to partly sacrifice yield of a cash crop to harvest the leaves and some fuel.
　　Recently <u>Eucalyptus hybrid</u> (Syn. <u>Eucalyptus teretecornis</u>) is being grown very extensively, mostly on field bunds in many parts of India. In this case also no economics of the system are known. It is reported that E. <u>hybrid</u> reduces yield of rainfed wheat (30) and E. <u>citriodora</u> reduces yield of many rainfed monsoon crops (e.g., <u>Phaseolus radiatus</u> and <u>Sorghum vulgare</u>).
　　Very recently a great amount of attention has been paid to <u>Leucaena leucocephala</u> and many benefits are being attributed to it. While it is being popularized on its reportedly good qualities, very little factual information is available about its interaction with agricultural crops or about economics of the system. When it is intercropped, its fodder and fuel yield tend to go down and intercrops yield less than pure crops. The type of crops grown and their yields are determined by the amount of rainfall. However, all this information is available only for 2-3 years and it is premature to draw valid conclusion at this stage (31, 42, 58).

Trees on farm boundaries and wood lots

<u>Farm boundaries</u>: Trees which are grown in the above settings are also usually grown on farm boundaries in all the agroecological regions of India. Apart from the specific uses to which these trees are put, they also help in demarcation of the farm and field boundaries, and serve as wind breaks and shelter belts. <u>Leucaena leucocephala</u> planted along field boundaries in Bundelkhand is reported to yield 3.21 tonnes fuelwood and 1.44 tonnes forage per ha 3 years after planting (53). Willows and poplars are grown on field boundaries in Lahaul in HP. <u>Delonix delta</u>, a green manure tree, is very popular with farmers in Tamilnadu for growing on field bunds and along boundaries of paddy fields; <u>Eugenia jambolana</u> and <u>Glyricidia spp.</u> are also grown for the same purpose. <u>Tectona grandis</u> is popular with the farmers of Tanjavur district in Tamilnadu. No information about the interaction of these trees with the agricultural crops or the economics of the system is available. <u>Prosopis juliflora</u> is observed to adversely affect the yield of agricultural crops in semi-arid areas when grown on farm boundaries. It is reported that yields of jowar

(<u>Sorghum vulgare</u>) were adversely affected as far as 30 m away from the tree (4 to 5 times the height of the trees) (57).

<u>Woodlots</u>: In many parts of India, farmers grow trees in separate wood lots. This system is expanding fast as the shortage of fuelwood is becoming more acute. <u>Casuarina equisetifolia</u> is extensively grown in AP, Tamilnadu and Karnataka states, on lands which are too poor or not very suitable for profitable agriculture. This has been practiced by the farmers of Tamilnadu much before India started to popularize farm forestry. The trees are planted very close and harvested in 5 to 7 years, when a yield of 120 tonnes fuelwood per ha may be obtained (63). <u>Casuarina</u> is clear-felled and stumps, which also secure a good price, are uprooted. Often the land is sequentially brought under agriculture for a year or two and then converted back to <u>Casuarina</u>.
 Farmers in Punjab have started diversifying their farming practices due to market economic pressures. It is now perceived by them that, due to acute shortage of fuel, as <u>Eucalyptus spp.</u> plantations could be commercially more viable than cropping. Farmers in Assam grow woodlots of bamboo along with paddy fields. There are some tree species which are grown only in specific conditions. Some examples are: 1. <u>Bambusa arundinacea</u> in depressed and water-logged areas in AP. These are harvested from the sixth year onwards and fetch very high prices. 2. In arid/semi arid Cuddapah district of AP, farmers grow <u>Pterocarpus santalinum</u> to get shafts for carts and plows. The tree is harvested in 12-15 years and coppices vigorously. 3. In the arid/semi-arid Anantpur district in AP, farmers do budding of "ber" (<u>Zizyphus spp.</u>) which grows wild in fields. The ber fruit gives high returns as compared to uncertain and poor crop yields (63).

Commercial crops under shade of trees

 Many commercial and plantation crops require shade for their optimum growth. Three types of practices can be recognized: (1) shade trees may be planted as a part of growing a commercial crop (e.g., tea, coffee); (2) a commercial crop (e.g., cacao) may be shaded by a planted commercial tree crop (e.g., coconut, arecanut); (3) shade may be provided by trees in a natural forest to a commercial crop (e.g., cardamom). The first two systems are agrosilviculture practices and the last one is a silvoagriculture practice. These are described together as a matter of convenience.

1. <u>Tea</u> (<u>Camellia sinensis</u>) is grown in large commercial plantations of 200 to 400 ha in Assam, WB and Tamilnadu states. It was recognized as early as 1863 that "Assam plant (<u>Camellia sinensis var assamica</u>) does not appear to thrive in inferior soils when exposed to full influence of the sun" and partial shading of tea bushes was recommended (36). Again as early as 1885, definite benefits of "sau" trees of Assam (<u>Albizia stipulata syn. A. chinensis</u>) were published (10). Legume trees (<u>Albizia spp., Acacia spp., Dalbergia serica</u> and <u>Derris robusta</u>) are used in northeast India and <u>Grevillea spp., Acacia spp.</u> and <u>Erythrina lithosperma</u> are used in south India. Shade could be provided by a single tree species or a mix. In tea gardens in Assam, low areas are given to laborers to grow paddy for their use. Tea shading, interactions between shade of different tree species, use of fertilizers, and yield of tea have been extensively investigated in India (16, 90).

Though shade trees have been planted in tea, their usefulness has always been a source of debate. A combination of observations, experience and experimentation has established that: (a) generally speaking, unshaded tea gives a lower yield than shaded tea; and (b) overly shaded tea gives a lower yield than medium or lightly shaded tea. From these observations it has become the practice to use shade trees which cause a light, even shade. It is important, therefore, to find a tree species or a mix that provides this. Leguminous shade trees benefit tea on two accounts, shade and nitrogen. The yield increment due to shade of <u>Albizia stipulata</u> is equivalent to 90 kg N per ha (90). Satisfactory crop yields can be obtained by the use of shade trees alone, but if shade trees are not used large quantities of nitrogen fertilizer are necessary (90). Tea is a mycorrhizal plant, and light intensity apparently is a determining factor for the formation of mycorrhiza. Thus it is suggested that certain nutrients become more available to the tea plant through its mycorrhizae as the light intensity is reduced (9).

It is customary in north eastern India to prune the tea bushes in winter. Weight of prunings is equivalent to the weight of tea leaves and shoots removed. Prunings are left on the ground and break down easily; usually by the end of monsoon they become incorporated into soil organic matter. It is reported that when the prunings of one season were removed there was a drastic reduction in the yield of tea (90). It is further reported that recovery of yield potential under this situation was better under <u>Albizia stipulata</u> than under bamboo screen shade. This may be explained by the fact that many leguminous trees, including <u>Albizia stipulata</u>, drop caducous stems every year, and these stems could account for much of the benefit (90).

2. <u>Coffee</u>: Coffee is cultivated commercially in the hilly tracts of the western and eastern ghats in the southern states of Karnataka, Tamilnadu, Kerala and Andhra Pradesh. Arabica (<u>Coffea arabica</u>) and robusta (<u>Coffee canephora</u>) are the two principle economic species extensively cultivated. In India, 90,211 ha (58%) are under arabica and 65,365 (42%) are under robusta. The total number of growers is 54,365 of whom the majority are classed as small growers, each with 10 ha or below (7).

The annual rainfall ranges from 1,250 to 3,000 mm, received mostly in the southwest monsoon. The plant grows well at temperatures between 12 and 36 degrees C. Arabica grows well at an elevation of 900 to 1,200 m and robusta grows at lower elevations (about 150 m). Both arabica and robusta have complex fertilizer schedules and doses depending on the age and expected yield of beans; fertilizers are applied in split doses depending on the growth stage of coffee in the season. Arabica is more shade tolerant than robusta in south India. Shade trees should be fast growing and spreading to allow a uniform amount of filtered light and their root systems should be deep so that they do not compete with coffee. The trees which serve as alternate hosts to the major pests of coffee must be avoided. The legume, dadap (<u>Erythrina lithosperma</u>), is almost universal as the temporary shade tree in India, planted with coffee in new clearings. The most popular permanent shade trees belong to the Indian fig and legume families, and include <u>Ficus glomerata</u>, F. <u>nervosa</u>, F. <u>tsjekela</u>, among the figs; and <u>Albizia stipulata</u>, A. <u>lebbek</u>, A. <u>moluccana</u> and <u>A. sumatrana</u> among the legumes. It is desirable to plant a large number of permanent shade trees at first and gradually thin them out as they grow and spread out. The trees have to be regulated in such a way that in course of time, they have their canopy about 10 to 14 m above coffee plants. Shade trees require constant attention by way of prunning and lopping to furnish the required shade (7).

3.   <u>Commercial cardamom</u> is of two types, "small" (<u>Elettaria cardamomum</u>) and "large" (<u>Ammomum</u> <u>subulatum</u> and <u>Ammomum</u> <u>aromaticum</u>).   Small cardamum cultivation is well organized.   In India the total area under cardamom is 93,947 ha, of which 60% is grown in Kerala, 30% in Karnataka and 10% in Tamilnadu states (11), almost all in small holdings.   About 50% of land is permanently owned by the grower and the rest is state government lease to the growers (24).   A mature cardamom plant may be 2 to 4 meters in height.   Cardamom generally grows well in loamy soils, with pH ranging from 5.0 to 6.5.   Adequate drainage is essential, and depending on the soil and local climatic conditions, it can be irrigated once during summer (11).   The natural habitat of cardamom is the evergreen forests of western ghats.   It is found between 600 and 1,200 meters elevation.   In most areas, annual rainfall is between 1,500 and 4,000 mm and the average temperature from 10 to 35 degrees C.

In the initial stages, cardamom plants require light shade for proper growth.   In areas having sparse tree growth, suitable quick growing shade trees are used.   <u>Mesopsis emini</u>, commonly known as African shade tree, is fast growing and can be grown as a temporary shade tree.   <u>Cedrella toona</u> or <u>Artocarpus heterophyllus</u> are good permanent shade trees.   Tall trees with well-distributed branching habit and small leaves are preferred.   A large number of nursery and plantation pests and diseases are recognized and various pest management practices recommended.   However, care must be used in applying chemicals, because bee management is common in plantations (11).

Trees are owned by the state governments, although growers can use dead, fallen trees for fuelwood.   Normal yield of small cardamon is 100 kg/ha, with economic yields into the 12th to 15th years.   Current research needs relate to developing high yielding and drought, pest and disease resistant varieties, effective control of pests and diseases, and developing suitable agro-techniques and improving post harvest techniques (24).

Large cardamom is cultivated in the north eastern Himalaya region in Sikkim, Darjeeling district of WB and Arunachal Pradesh.   These systems have not been described in detail.   Large cardamom is grown under the shade of natural forest as well as on flat or sloping lands by the planting of shade trees.   Large cardamom takes 4 to 5 years to come to production and is reported to yield 250-300 kg of cardamom per ha (22, 23, 86).

8.4.3. <u>Pastoralsilviculture Systems</u>

Indian farmers have integrated crop farming with very large numbers of livestock (43).   These livestock depend on grasses and fodder for their survival.

Most of the livestock graze freely in village grazing lands and adjoining forest lands, where the villagers have rights to graze their animals.   Livestock also graze at large in agricultural fields when no crops are growing.   (This is very well regulated by traditional and social practice.)   The agroecological and edaphic conditions in India lead to growth of trees even in grasslands (except in alpine pastures); conversely grass will also grow in most forests, except in tropical high rainfall areas where the forest canopy is closed.

The practices that have grazing as the major component and a scattering of trees are referred to as pastoral-silviculture systems in this chapter. These are practiced extensively by most all farmers in the country with the types of grasses and trees varying with the local conditions. Some of the practices are very well-developed.

Grassland and tree management in the semi-arid zone

This system is practiced in AP, Karnataka, Tamilnadu, Maharashtra and Madhya Pradesh in the Deccan plateau at an elevation of 300-1000 m. Average annual rainfall varies from 500-1300 mm in 45-70 rainy days occurring mainly during southwest monsoon. This system is locally called "kanchas". The kanchas are owned by the cultivators, although some are with the Government Forest Departments.

In this case, the land with existing trees is left fallow and protected. Natural succession of the grasses takes place. The grass crop is protected for one to three years depending on the economic condition of the farmer. The grasses fall mainly in the Sehemia-Dicanthium grassland type, grow fast and can reach a mature stage in 4-10 years depending on location and site conditions (13).

Many trees are represented in this climate. The farmer determines which trees may be encouraged to grow or be planted. Commonly planted trees are Eucalyptus hybrid and Casuarina equisetifolia. Palms like Borassus flabellifer and Phoenix sylvastus are common. Trees are lopped for fodder and fuel. Custard apple, mango, zizyphus and tamarind fruits are collected for home consumption. Neem (Azadirachta indica) fruits are collected for sale; mahua (Madhuca latifolia) fruits yield edible oil and its flowers are used to brew alcoholic drink.

Animals are an integral part of the practice; grasslands are grazed and tree fodder lopped.

Resource input is restricted to enclosing the area, planting grass or trees and managing them. Yields of grass depend on the stage of succession and management. Since farmers provide 1-2 years protection, yields are only 20 to 50% of their potential if the area is protected for at least four years. In a well protected kancha on red soils, 3 tonnes air dry grass per ha per annum may be obtained; in gravelly soils the yield is 0.5 to 1.0 air dry grass per ha per annum; a well-protected kancha on black soil can yield 10 tonnes (13).

Kanchas are an ideal production and protective system in semi-arid lands. Due to lack of vegetative cover as well as intensive rainfall, the lands are potentially exposed to erosion. The grass and tree cover, if managed well, prevents soil erosion. The grass yield is certain while crop yield is uncertain and is also accompanied by erosion hazard.

Grassland and tree management in arid zone

The importance of Prosopis cineraria and Zizyphus nummularia in agrosilviculture systems in the hot arid zone has been described earlier in this chapter. These two species play a vital role in the pastoralsilviculture system also. The community and village grazing lands in Rajasthan are called "oran" or "bir". Both these species are encouraged and managed in the orans.

Prosopis cineraria, when cut above the ground level, produces numerous new buds which give rise to several (5-12) new shoots. These in a year's time assume a bushy structure. The coppiced shoots are continuously grazed in an oran, leading to a "cushion" form of the crown. Under severe grazing and trampling the newly sprouted branches spread horizontally and provide feed to small animals like sheep (67).

The shrubs of _Zizyphus nummularia_ have remarkable powers of regeneration through root suckers. It is reported that 14% density of _Z. nummularia_ bushes is optimal for obtaining the maximum yield of leaf fodder and grass (28). As in the case of agricultural crops, _Prosopis cineraria_ is very compatible with grass (73). Apart from leaf fodder, mature _Zizyphus_ yields about 3.0-5.0 kg of fruit per bush growing in an oran in a 150-250 mm rainfall zone (67). (It does not yield fruits in agricultural lands, as it is cut every year.)

Other grass and tree management practices

Closure to biotic interferences: Prevention of grazing, browsing by other animals, lopping, etc., often restores the indigenous grass and tree cover. Once the pressure is taken off, native grasses are re-established in a very short time. For example (84, 85), in the alluvial ravines in Gujarat as a result of closure to grazing, _Aristida funiculata_ and _Themada triandra_ were replaced first by _Apluda mutica_ (an annual grass) and then by perennials like _Eremopogon faveolotus_, _Heteropogon contortus_, _Dicanthium annulatum_ and _Cenchrus_ spp. There was not only a qualitative change in species but there was a quantitative increase in the yield of grasses. The number and size of trees also increased over time (55). Runoff and soil erosion losses progressively decreased from the area. Similar results are reported from semi-arid alluvial ravines on the banks of Jamuna river at Agra (56), ravines on the banks of Chambal river (78, 79), and the arid zone of Rajasthan (8).

Plantings: Experiments have been conducted to replant low quality lands with desirable grass and tree species suited to the prevailing conditions. These experiments were initiated in the late 1950's and were called "fodder-fuel plantations". In the alluvial ravine lands in Gujarat, _Cenchrus ciliaris_ and _Dicanthium annulatum_ were compatible with _Acacia nilotica_. The overall mean yield of grasses increased from year to year, reaching a maximum of 101% increase in the fourth year (17). As the canopy of the trees increased, the grass yield should decrease. In black clayey soil with similar annual rainfall (at Kota), there is evidence that in association with _Acacia nilotica_, there was not only less yield of grass as compared to alluvial soils, but the grass yield was also reduced at an earlier stage (4 years after planting) as the trees grew (14). On semi-arid black soil, _Cenchrus ciliaris_ gives much higher yield (3.4 tonnes green grass/ha) than _Dicanthium annulatum_ (1.7 tonnes green grass/ha) in association with _Acacia nilitica_ trees over a period of six years (83).

In the high rainfall, sub-tropical humid climate at Dehradun, when _Chrysopogon fulvus_ grass and _Dalbergia sissoo_ trees were planted together on bouldary river terraces, a very high yield of grass (10.55 tonnes/ha) was obtained and 97 trees/ha were available for harvesting for fodder and fuel (37). In another experiment, under similar edaphic conditions at Dehradun, over a period of 4 years when the trees (_Albizia lebbek_, _Grewia optiva_, _Bauhinia purpurea_, and _Leucaena leucocephala_) were establishing themselves, an average grass (_Chrysopogon fulvus_ and _Eulaliopsis binata_) yield of 3.9 to 4.4 tonnes (oven dry) per ha was obtained per cutting. _Eulaliopsis binata_ (an industrial grass used for rope making and paper

pulp) gave a higher yield (5.4 tonnes oven dry grass per ha) than
Chrysopogon fulvus (3.3 tonnes per ha) per cut. All the tree species had
better growth in association with Chyrsopogon than with Eulaliopsis (80).

8.4.4. Silvopastoral Systems

These systems involve lopping trees and grazing understory grasses
and bushes in forests and plantations. While human and livestock
populations were within reasonable limits, forests were able to supply
their needs without degradation. Now all types of forests and grasslands
are severely degraded due to over exploitation. Yet surprisingly little
specific facts and figures have been reported.

One of the more outstanding examples of silvopastoralism is grazing
in forest lands in Himalaya. In central and western Himalaya, in the
states of Jammu and Kashmir (J&K), HP and UP, the forests are divided into
tropical, sub-tropical and temperate zones (the alpine zone does not have
forests). In summer the animals are grazed in the alpine pastures and as
the cold weather comes, the animals are moved down to temperate forests
and finally to sub-tropical forests. The pastures are reported to be over
grazed though no quantitative data are available.

The temperate zone supports forests of Cedrus deodora, Picea
smithiana, Abies pindrow and various pines. Along the streams and
depressions broad-leaved species, like Acer, Cornus, Betula, and Quercus,
grow. Migratory flocks of sheep, goat and cattle pass through these
forests and cause heavy damage to the grass and tree regeneration. This
zone is inhabited, so that animals of farming communities also contribute
to the problem of over grazing and over lopping of trees in the forests.
The sub-tropical zone is relatively more densely populated and the
pressure is much more there. The dominant forest tree species are Shorea
robusta, Acacia catechu, Pinus roxburghii and Quercus incana. The extent
of degradation of the forest in a sub-tropical to temperate watershed of
370 ha can be gauged by the fact that out of 133 ha of forest land, 26% of
the forest land had no canopy, 12% had a thin canopy, and 62% had a
moderately dense canopy; there was no dense forest at all (5). This is
not an isolated example, but is widely true of the western and central
Himalaya.

Experiments have been conducted to grow grasses along with densely
planted Eucalyptus spp. Eucalyptus does not encourage growth of grasses
in low rainfall areas, but in areas having more than 1000 mm of rainfall,
preferably in monsoon as well as in winter, some grasses grow under it.
In this case the goal has been to create fuel tree plantations as well as
obtain some grass yield as a bonus. The situation is similar to
intercropping under coconut or with horticultural trees. In one
experiment (40, 41), Eulaliopsis binata grass and tall plants (2m and
above) of Eucalyptus hybrid were planted at a spacing of 2m x 2m. To
promote moisture conservation tie ridges, 15 cm high at 2m x 2m spacing
were also made. Within a period of 8 years the trees had attained an
average height and girth of 14.7 m and 38.2 cm, respectively. During a
period of 8 years 45.2 tonnes of air dry grass/ha, and during a period of
8 1/2 years 179 tonnes of air dry tree biomass/ha were obtained (this
comprised 148 tonnes of poles, 19 tonnes of fuel, and 12 tonnes of leaf).
When the grass was grown alone it yielded 114.4 tonnes air dry grass
during a period of eight years. After the Eucalyptus plot was harvested,
the annual yield of grass increased and 9.8 tonnes of air dry grass per ha
was obtained in the Eucalyptus plot as compared to 13.6 tonnes in the

128

grass alone plot. In the second year after coppicing, the _Eucalyptus_ plants grew very vigorously and coppice shoots attained a height of 7.0 m and a girth at breast height of 14.9 cm. The grass yield in _Eucalyptus_ plots had decreased to 4.25 tonnes, reflecting the interaction with shade of coppice shoots. In another experiment at Dehradun (59), _Chrysopogon fulvus_ fodder grass was grown in association with _Eucalyptus_ hybrid. It was observed that over a period of four years, air dry grass yields of 3.9, 3.4 and 3.9 tonnes/ha/year were obtained when _Eucalyptus_ was grown at a spacing of 3.5 m x 3.5 m, 5.0 m x 5.0 m and 7.0 m x 7.0 m, respectively.

8.4.5. Agrosilvopastoral Systems

Since Indian farmers usually have integrated animals fully with their farming operations, many of the agroforestry systems are truly agrosilvopastoral systems. Some examples include intercropping with coconut and horticultural trees if grasses are also grown as intercrops, trees in agricultural fields and trees on farm boundaries. The animal component is variable (e.g., the systems of trees in agricultural fields and farm boundaries will have relatively more dependence on animals than the systems of growing agricultural crops with commercial trees). Even in shifting cultivation, there is an animal component of pigs and poultry.

Some of the silvopastoral systems would also qualify agrosilvopastoral systems when they interface with agriculture. However, none of these systems has been described since the animal component has never been evaluated. However, home gardens truly combine animals, birds and fish with crops and trees (89).

Home Gardens

The system of home gardens is practiced extensively in high rainfall areas in tropical south and southeast Asia. This practice finds expression in the states of Kerala and Tamilnadu, with humid tropical climates and where coconut is the main crop. Only very recently have home gardens in India been described (48). Edaphic conditions are similar to those for coconut and arecanut, already described in this chapter.

Most home gardens also support a variety of animals (cows, buffaloes, bullocks, goats, sheep) and birds (chicken, duck). In certain places pigs are also raised. Fodder and legumes are widely grown to meet daily fodder requirements of cattle. The waste materials from crops and homes are used as fodder/feed for animals/birds and barn wastes are used as manure for crops (48).

Mangroves form an essential part of the homesteads of backwater areas in the lowlands and coastal tracts. _Acanthus illicifolius_, _Avicennia officinalis_, _Carbera odollam_, _Rhizophora conjugata_ and _R. mucronata_ are common. Mangroves form an ideal habitat for fish farming, with detritus serving as feed for fish. Fish and prawns are also grown in farm ponds, paddy fields and canals (48), and some home gardens also maintain bee hives.

Wide variation in the intensity of tree cropping is noticeable between home gardens situated in the same agroclimatic zone, and a reduction in the size of the holding leads to an intensification of cultivation (46). It is commonly noticed that, as the intensity of tree cropping is increased, miscellaneous tree species having no immediate benefit are replaced with multiple use species. Although this reduces species diversity, the intensity of cultivation/use of selected species increases considerably. Variations in the intensity of tree cropping are also influenced by differences in socio-economic conditions of the farmers and their response to externally determined changes, particularly prices of inputs and products, variations in land tenure systems, etc. (46, 48).

Although it seems that the trees and crops are not grown in any specific pattern, it is difficult to accept that a system which has evolved over a long time could be casual about location, spacing and site conditions of important crops. It may be assumed that those who practice it know in a practical way what, where and when to plant and remove.

An average household consists of 6-8 persons who provide the workforce for the home garden. As compared to monocropping, the home gardens involve high labour input. For example, a home garden with an intensive crop mix may require 1000 days per year as compared to 150 days for coconut monocropping and 400 days for rice monocropping (48).

Seeds are usually stored by the farmers; seeds/seedlings are also obtained from authorized government agencies and private suppliers. Chemical fertilizers are being increasingly used but plant protection chemicals are not widely used.

Information regarding the relative yields of agricultural and tree crops and animals in individual home gardens is not available. Nair and Sreedharan (48) have only reported average production per unit area or plant, and this does not help in evaluating the benefits of home gardens.

Economic yields have been obtained in one case study of a 0.12 ha home garden (29). Although the values are in the realm of speculation, these results and others obtained outside India indicate that the system is remunerative and would provide a good subsistence.

The major constraint of the system is that it is the least understood scientifically. The improvement of the system is very challenging and potentially very promising. Questions related to resource utilization, use of high yielding, short duration, fast-growing varieties, spatial and temporal arrangements of plants, long-term effects on the stability of the production base, protective benefits and input-output relations need addressing.

8.5. CONCLUSIONS

Although agroforestry practices in India are very old and traditional, they are currently not well understood or even described. There are a number of basic issues which have to be understood before it is possible to improve these systems. These include the interactions between the trees and crops/grasses with respect to shade, rooting patterns, competition for plant nutrients and moisture, soil fertility, and compatibility between various tree and crop/grass species. Applied and basic research should be conducted side by side. Is it possible to select fast-growing, high yielding, more compatible tree species native to India? Is it possible to arrange better spacings to obtain higher sustainable production per unit of space, unit of time and unit of water? Is it possible to prepare models to predict responses?

Indian farmers (as farmers all over the world) have developed most of these systems in response to their subsistance needs and the agroecological situations in which they live. When the research results are critically examined, it seems as if scientists so far have not materially improved the productivity of any of these systems. Clearly, this is where the challenge and the opportunities await.

ACKNOWLEDGEMENTS

To so briefly write about a vast subject like agroforestry systems
and research in India is a formidable task. I am grateful to the Indian
Council of Agricultural Research, International Council for Research in
Agroforestry, and the East-West Center for providing opportunities for
work and to the many colleagues with whom I have interacted and have had
disucssions on the subject. Particular mention may be made of Bjorn
Lundgren, P. K. R. Nair, Napoleon Vergara, K. A. Shankarnarayan, K. V. A.
Bavappa, S. Chinnamani, Ch. Ram Prakash and S. S. Teotia. More than 200
references were consulted/documented for preparation of this chapter. As
the limitations on the size of the chapter would not permit citing all of
the individual papers both published and unpublished, I take this
opportunity to thank all the organizations and individuals whose works
have not been referenced.
Finally, I am thankful to the International Center for Integrated
Mountain Development, Kathmandu, Nepal and its Director, Colin Rosser, for
the institutional support provided to me.

REFERENCES

1.   Acharya, R. M., Patnayak, B. C. and L. D. Ahuja. 1977. Livestock
     production problems and prospects. In: Desertification and its
     control, pp 275-280. Indian Council Agric Res, New Delhi, India.
2.   Aggarwal, R. K. 1980. Physiochemical status of soils under Khejri
     (Prosopis cineraria Linn) In: H. S. Mann and S. K. Saxena (eds)
     Khejri (Prosopis cineraria) in the Indian desert, pp 32-37. Central
     Arid Zone Res Inst, Jodhpur, India.
3.   Anonymous. 1979. Package of practices for arecanut. Pamphlet no 2E,
     Central Plantation Crops Res Inst, Kasargod, India.
4.   Anonymous. 1981a. Proceedings of the summer institute on agroforestry
     in semi-arid zones. Central Arid Zone Res Inst, Jodhpur, India.
5.   Anonymous. 1981b. Report on operational research project: Watershed
     management. Central Soil Water Cons Res Training Inst, Dehra Dun,
     India.
6.   Anonymous. 1982. Shifting cultivation in north eastern region. North
     Eastern Council Pub No 17, pp 1-24, Shillong, India.
7.   Bavappa, K. V. A. 1980. Plantation crops-Arecanut. In: Handbook of
     Agriculture, pp 865-920. Indian Council Agric Res, New Delhi, India.
8.   Bhimaya, C. P., Kaul, R. N. and L. D. Ahuja. 1967. Forage production
     and utilization in arid and semi-arid rangelands. Proc Symp Sci
     India's Food, pp 215-223. Nat Inst Sci, and Indian Council Agric Res,
     New Delhi, India.
9.   Bjorkman, E. 1942. Symbolae botanicae Upsaliensis. 6:1-190. (Quoted
     from Biol Absts 20:7878 (1946) by Wight, 1959).
10. Buckingham, J. 1885. Papers regarding the sau tree and its remarkable
     influence on tea bush. Indian Tea Assoc, Calcutta (cited from Wight,
     1959).
11. Cardamom Board. 1986. Cardamom cultivation. Cardamom Board, Cochin,
     India. 38 pp.
12. Census of India. 1982. Primary census abstract. General population
     series 1, Part II B(I). Census Commission India. Controller
     Publications, Delhi, India.
13. Chinnamani, S. 1983. Pers Comm.

14. Chinnamani, S., Rathore, B. L. and M. C. Prajapati. 1982. Scientific management and production potential of tree and grass species in large scale plantations on sub-catchment basis in Chambal ravines at Kota, pp 84-87. Ann Rep, Central Soil Water Cons Res Training Inst, Dehra Dun, India.

15. Combe, J. 1982. Agroforestry techniques in tropical countries: potential and limitations. Agrofor Syst 1:13-28.

16. Cooper, H. R. 1939. Nitrogen supply to tea. Indian Tea Assoc, Calcutta, India (cited from Wight, 1959).

17. Dayal, R. 1974. Report on achievements of soil conservation research, demonstration and training center, Vasad, India. (Unpub).

18. Gangolly, S. R., Singh, R., Katyal, S. L. and D. Singh. 1957. The mango. Indian Council Agric Res, New Delhi, India. 530 pp.

19. Ganguli, B. N., Kani, R. N. and K. T. N. Nambiar. 1964. Preliminary studies on a few crop feed species. Annals Arid Zone 3:33-37.

20. Ghildyal, B. P. 1981. Soils of the Garhwal and Kumaon Himalayas. In: J. S. Lal (ed) The Himalaya: Aspects of change, pp 120-151. Oxford Univ Press, Delhi, India.

21. Gupta, B. L. 1956. Forest flora of Chakrata, Dehradun and Saharanpur forest divisions. Third Edition. Manager of Publications, Gov of India, Delhi, India. 558 pp.

22. Gupta, P. N. 1982. Export potential in large cardamom. Indian Hort 26:21-25.

23. Gupta, P. N. 1983. Cardamom in Sikkim state (pers comm).

24. ICRI. 1987. Indian Cardamom Res Inst, Myladumpara, India. (Unpub).

25. Jambulingam, R. and E. C. M. Fernandes. 1986. Mulitpurpose trees and shrubs on farmlands in Tamilnadu state (India). Agrofor Syst 4:17-32.

26. Jodha, N. S. 1977. Land tenure problems and policies in the arid region of Rajasthan. In: Desertification and its control, pp 335-347. Indian Council Agric Res, New Delhi, India.

27. Kannan, R. and K. P. P. Nambiar. 1976. Studies on intercropping coconut garden with annual crops. Coconut Bull 5:1-3.

28. Kaul, R. N. and B. N. Ganguli. 1963. Fodder potential of Zizyphus in the scrub grazing lands of arid zone. Indian For 89:623-630.

29. KGSN. 1984. Balanced (small) farm - an appropriate garden technology. Kerala Gandhi Samarak Nidhi, Trivandrum (pamphlet cited from Nair and Sreedharan 1986).

30. Khybri, M. L. and Sewa Ram. 1984. Effect of Grewia optiva, Morus alba and Eucalyptus hybrid on the yield of crops under rainfed conditions. pp 51-53, Ann Rep, Central Soil Water Cons Res Training Inst, Dehra Dun, India.

31. Khybri, M. L., Sewa Ram and S. P. Bhardwaj. 1984. Study on intercropping of field crops with fodder crop of "subabul" (Leucaena leucocephala) under rainfed conditions. pp 61-62, Ann Rep, Central Soil Water Cons Res and Training Inst, Dehra Dun, India.

32. Kushwah, B. L., Nelliat, E. V., Markose, V. T. and A. F. Sunny. 1973. Rooting pattern of coconut (Cocos nucifera L.). Indian J Agron 18:71-74.

33. Malhotra, S. P. 1977. Socio-demographic factors and nomadism in the arid zone. In: Desertification and its control, pp 310-323. Indian Council Agric Res, New Delhi, India.

34. Mann, H. S., Malhotra, S. P. and K. A. Shankarnarayan. 1977. Land and resource utilization in the arid zone. In: Desertification and its control, pp. 89-101. Indian Council Agric Res, New Delhi, India.

35. Mann, H. S. and S. K. Saxena. 1980. Khejri (Prosopis cineraria) in the Indian desert. Monograph No 11, Central Arid Zone Res Inst, Jodhpur, India. 78 pp.

36. Masters, J. W. 1939. J Agric Hort Soc India 13:31-47 (cited from Wight, 1959).

37. Mathur, H. N. and P. Joshie. 1972. Ann Rep, Central Soil Water Cons Res Training Inst, Dehra Dun, India. (mimeo).

38. Menon, K. P. V. and K. M. Pandalai. 1960. The coconut palm-a monograph. Indian Central Coconut Committee, Ernakulam, India.

39. Mishra, B. K. and P. S. Ramakrishnan. 1981. The economic yield and energy efficiency of hill agro-ecosystems at higher elevations of Meghalaya in north eastern India. Acta Oecologica Oecol Applic 2:369-389.

40. Mishra, P. R. and A. D. Sud. 1983. pp 57-60, Ann Rep, Central Soil Water Cons Res Training Inst, Dehra Dun, India.

41. Mishra, P. R. and A. D. Sud. 1984. Tie ridge technique of raising Eucalyptus and bhaber on 4% slope. pp 84-85, Ann Rep, Central Soil Water Cons Res Training Inst, Dehra Dun, India.

42. Mittal, S. P. and P. Singh. 1983. Study on intercropping of field crops with fodder crop of "subabul" (Leucaena leucocephala) under rainfed conditions. Ann Rep, Central Soil Water Cons Res Training Inst, Dehra Dun, India.

43. MOA. 1982. Indian agriculture in brief. Ministry Agric, New Delhi, India.

44. MOA. 1983. State wise area and production of important fruits in India from 1976-77 to 1978-79 and projections up to 1982-83. Ministry Agric, New Delhi, India. (unpublished).

45. Muthana, K. D. 1980. Silviculture aspects of Khejri. In: H. S. Mann and S. K. Saxena (eds) Khejri (Prosopis cineraria) in the Indian desert, pp 20-24. Central Arid Zone Res Inst, Jodhpur, India.

46. Nair, C. T. S. and C. N. Krishnankutti. 1984. Socio-economic factors influencing farm forestry. A case study of tree cropping in homesteads in Kerala, India. In: Y. S. Rao, N. T. Vergara and G. W. Lovelace (eds) Community forestry: Socio-economic aspects, pp 115-130. FAO Regional Office Asia-Pacific Region, Bangkok, Thailand.

47. Nair, K. K. 1981. Agroforestry in the southern hills. In: Proc Agrofor Seminar, pp 234-241. Indian Council Agric Res, New Delhi.

48. Nair, M. A. and C. Sreedharan. 1986. Agroforestry farming systems in the homesteads of Kerala, southern India. Agrofor Syst 4:339-363.

49. Nair, P. K. R. 1979. Intensive multiple cropping with coconuts in India. Verlang Paul Parey, Berlin and Hamburg.

50. Nair, P. K. R. 1985. Classification of agroforestry systems. Agrofor Syst 3:97-128.

51. NCA. 1976. Report of the National Commission of Agriculture. Part IX, Abridged Rep, pp 427-472. Gov of India, New Delhi, India.

52. Nelliat, E. V. and K. S. Bhat. 1979. Multiple cropping in coconut and arecanut gardens. Tech Bull 3, Central Plantation Crops Res Inst, Kasargod, India.

53. Patil, B. D., Deb Roy, R. and P. S. Pathak. 1981. Agroforestry research and development with reference to Indo-Gangetic plains. In: Proc Agrofor Seminar, pp 40-68. Indian Council Agric Res, New Delhi, India.

54. Planning Commission. 1982. Socio-economic development of Himalayan hills. Task force for the study of ecodevelopment of Himalayan Region. Gov of India, New Delhi, India.

55. Pradhan, I. P. 1977. Annual Report. Central Soil Water Cons Res Training Inst, Dehra Dun, India (Mimeo).

56. Prajapati, M. C. 1974. Report on achievements (1955-74) of Soil Conservation Research, Demonstration and Training Center, Agra. (Mimeo, M. C. Prajapati and S. S. Sajwan, pers comm).

57. Prajapati, M. C. et al. 1971. Effect of lateral development of Prosopis juliflora D. C. roots on agricultural crops. Annals Arid Zone 10(2&3):186-193.

58. Prasad, S. N. and B. Verma. 1983. Intercropping of field crops with fodder crop of subabul (Leucaena leucocephala) under rainfed conditions. pp 95-96, Ann Rep, Central Soil Water Cons Res Training Inst, Dehra Dun, India.

59. Puri, D. N. and P. Joshie. 1979. Economic utilization of class V, VI and VII lands for raising fuel (Eucalyptus hybrid) and fodder (Chrysopogon fulvus). Ann Rep, Central Soil Water Cons Res Training Inst, Dehra Dun, India.

60. Purohit, M. L. and Wajid Khan. 1980. Socio-economic dimensions of Khejri. In: H. S. Mann and S. K. Saxena (eds) Khejri (Prosopis cineraria) in the Indian desert, pp 56-63. Central Arid Zone Res Inst, Jodhpur, India.

61. Ram Prakash. 1983. Pers Comm.

62. Ramakrishnan, P. S. and O. P. Toky. 1978. Preliminary observations on the impact of shifting agriculture on the forested ecosystem. In: Nat Seminar Resour Develop Envir Himalayan Region, pp 343-354. Gov India, New Delhi, India.

63. Reddy, C. V. K. 1981. Agroforestry in coastal Andhra Pradesh. In: Proc Agrofor Seminar, pp 77-82. Indian Council Agric Res, New Delhi.

64. Sahu, S. B., Singh, A. and M. D. Singh. 1980. Farm power sources and their utilization in north eastern hill region. Res Bull 7, Indian Council Agric Res, Res Complex, Shillong, India. 22 pp.

65. Samraj, P., Haldorai, B. and C. Henry. 1982. Economic utilization by intercropping in afforested areas. pp 47-49, Ann Rep, Central Soil Water Cons Res Training Inst, Dehra Dun, India.

66. Saxena, S. K. 1980. Taxonomy, morphology, growth and reproduction of Khejri and its succession in northwest India. In: H. S. Mann and S. K. Saxena (eds) Khejri (Prosopis cineraria) in the Indian desert, pp 4-10. Central Arid Zone Res Inst, Jodhpur, India.

67. Saxena, S. K. 1984. Khejri (Prosopis cineraria and bordi (Zizyphus nummularia): multipurpose plants of arid and semi-arid zones. In: K. A. Shankarnarayan (ed) Agroforestry in arid and semi-arid lands, pp 111-121. Pub No 24, Central Arid Zone Res Inst, Jodhpur, India.

68. Seshachalam, N. 1979. pers comm (cited from Tejwani, K. G. 1981. Agroforestry for the wasted lands in western Himalayas. In: Proc Agrofor Seminar, pp 220-233. Indian Council Agric Res, New Delhi, India).

69. Seth, S. K. 1981. India and Sri Lanka: Agroforestry. Report for local community development programme. UN/Food Agric Org (FAO), Rome, Italy.

70. Shah, S. L. 1981. Agricultural planning and development in the northwestern Himalayas, India. In: Nepal's experience in hill agricultural development, pp 160-168. Ministry Food Agric, Kathmandu, Nepal.

71. Shankar. V. 1980a. Khejri (Prosopis cineraria) in Indian scriptures. In: H. S. Mann and S. K. Sazena (eds) Khejri (Prosopis cineraria) in the Indian desert, pp 1-3. Central Arid Zone Res Inst, Jodhpur, India.

72. Shankar, V. 1980b. Distribution of Khejri (Prosopis cineraria Macbride) in western Rajasthan. In: H. S. Mann and S. K. Sazena (eds) Khejri (Prosopis cineraria) in the Indian desert, pp 11-19. Central Arid Zone Res Inst, Jodhpur, India.

73. Shankar, V. 1984. Inter-relationships of tree overstorey and understorey vegetation in silvipastoral system. In: K. A. Shankarnarayan (ed) Agroforestry in arid and semi-arid lands, pp 143-149. Central Arid Zone Res Inst, Jodhpur, India.

74. Shankarnarayan, K. A. (ed) 1984. Agroforestry in arid and semi-arid lands. Central Arid Zone Res Inst, Jodhpur, India. 293 pp.

75. Sharma, T. C. 1980. The prehistoric background of shifting cultivation. In: Shifting cultivation in northeast India, pp 1-5. Northeast India Council Social Sci Res, Shillong, India.

76. Singh, A. 1981. Shifting cultivation. In: A. N. Asthana, S. P. Ghosh, P. S. R. C. Murti and R. N. Verma (eds) Proc Workshop Agric Res in North eastern Hills Region, pp 174-178. Indian Council Agric Res, Res Complex, Shillong, India.

77. Singh, A. and M. D. Singh. 1981. Soil erosion hazards in northeastern hill region. Bull No 10, Indian Council Agric Res, Res Complex, North eastern Hill Region, Shillong, India. 30 pp.

78. Singh, B. 1971. A comparative study on economics of various soil conservation cum grassland improvement practices in ravine lands, II. Indian For 97:387-391.

79. Singh, B. and B. Verma. 1971. A comparative study on economics of various soil conservation cum grassland improvement practices for rejuvinating forage production in ravine lands, I (forage production). Indian For 97:315-321.

80. Singh, H. and Joshie, P. 1984. Economic utilization of bouldry riverbed land for fodder, fibre and fuel under rainfed conditions. pp 57-60, Ann Rep, Central Soil Water Cons Res Training Inst, Dehra Dun, India.

81. Singh, R. K., Yadukumar, N., RoyBurman, K. N. and A. C. Roy. 1982. Intercropping in arecanut gardens in north Bengal. Indian Farming 32:13-15.

82. Singh, S. Krishnamurthi, S. and S. L. Katyal. 1963. Fruit cultivation in India. Indian Council Agric Res, New Delhi. 451 pp.

83. Subbayan, R. and S. Chittaranjan. 1979. Fodder-cum-fuel plantation in the black cotton soils of Bellary under rainfed conditions. pp 57-58, Ann Rep, Central Soil Water Cons Res Training Inst, Dehra Dun, India.

84. Tejwani, K. G., Dhruvananyana, V. V. and T. Satyanarayana. 1960. Control of gully erosion in ravine lands of Gujarat. J Soil Water Cons (India) 8:65-74.

85. Tejwani, K. G., Srinivasan, V. and M. S. Mistry. 1961. Gujarat can save its ravine land. India Farming 11:20-21.

86. Uppadhaya, R. C. and S. P. Ghosh. 1983. Wild cardamom of Arunachal Pradesh. Indian Hort 27:25-27.

87. Varghese, P. T. et al. 1978a. Intercropping with tube crops in coconut gardens. In: Proc Placrosym, pp 399-405. 1. Placrosym Standing Committee, Kasargod, India.

88. Varghese, P. T. et al. 1978b. Beneficial interactions of coconut cacao crop combination. ibid, pp 383-392.

89. Wiersum, K. F. 1982. Tree garden and taungya on Java. Examples of agroforestry in the humid tropics. Agrofor Syst 1:53-67.
90. Wight, W. 1959. The shade tree tradition in the tea gardens of north India: I. The value of shade. pp 75-98, Rep Indian Tea Assoc, Sci Dep Tocklai Expt Sta, Jorehat, India.

9. DEVELOPMENT OF TAUNGYA IN NIGERIA

R. G. Lowe

Department of Forest Resources Management, University of Ibadan, Nigeria

9.1. ABSTRACT

'Taungya' is a Burmese word, applied in tropical silviculture to the adaptation of shifting cultivation whereby natural tropical forest is replaced by forestry plantations, with the intention of increasing timber productivity. The method was introduced to Nigeria in 1926, to enable forestry workers and their families to feed themselves, and then used as a means to conduct experimental tree planting; it was subsequently extended into the pilot phase of forestry plantations, sometimes by means of deliberately established Taungya settlements. It is essentially a system of shifting cultivation, organized by the Forestry Department within forest reserves, and depends on the existence of a peasant population in a largely subsistence economy, at a low standard of living.

After the end of the Second World War, the high forest reserves in the south of Nigeria became progressively exploited for timber and a method of regeneration was sought that could keep up with the rate of exploitation. This led in the 1950's to a preference for the Tropical Shelterwood System of natural regeneration. However, in the 1960's, the forest reserves came under pressure from alternative forms of land use, such as commercial plantation crops of cocoa, oil palm or rubber, as well as from peasant arable cropping - and something more than natural regeneration had to be employed if the land were to remain in forestry use. Therefore forestry activity reverted to Taungya plantations in accessible forest.

In the 1970's, after the Civil War, the oil-boom years led to enhanced economic development, and to the country itself consuming the timber that had formerly supplied export markets; so that in 1976 all export of timber was forbidden. It is forecast that by the year 2000, the country's timber resources will become exhausted. Some forest reserves are already being exploited for the third time since 1945. If the deficit is to be met, an estimated 100,000 hectares of timber crop plantations must be established annually. However, various factors have made people, especially young people, unwilling to become engaged in peasant agriculture and condemned to a relatively low standard of living. The only alternative has been for the Forestry Department to plant the arable crops itself - employing wage labor for the work - as well as establishing the tree crop.

In the 1980's, pulpwood projects financed by the World Bank have turned to mechanized land clearing and cultivation. It has also become apparent that the State forestry departments are incapable of the large-scale commercial planting now needed. In fact, some extensive areas of forest reserve are being laid waste by arable cropping without a tree crop being established. State governments have more pressing needs than forestry on which to spend their revenues. Nevertheless, the economic

arguments for a combination of arable cropping with tree crop
establishment are cogent, and this is more profitable than either alone.

9.2. INTRODUCTION

'Taungya' is a part of the history of forestry in Nigeria, and indeed
of the country itself. In 1897 Cyril Punch was appointed Superintendent
in the newly established Department of Woods and Forests of the Colony and
Protectorate of Lagos, which itself had been proclaimed in 1886; in 1899
Peter Hitchen was recruited as Inspector of Forests, Southern Nigeria,
with his headquarters at Calabar, although he resided mainly at Benin
City, which neighbored the largest and most accessible block of forest in
the country. In the same year, the Mamu Forest Reserve was created,
incidentally forming a buffer between the territories of the city states,
Ibadan and Ijebu; and in 1900 the Olokemeji Forest Reserve was also
constituted between Abeokuta and Ibadan. In 1901, the 1st Forestry
Ordinance was promulgated, to regulate the size of timber concessions, to
impose forestry fees based on the area of the concession and on each tree
felled, and to oblige concessionaires to plant 20 economic tree seedlings
at each stump site. Duties were also exacted on all exported timber.
Moreover the ordinance declared that "All forest lands at the disposal of
Government upon which timber or rubber is now being cut or collected" were
forest reserves. An important activity of the Forestry Department was
that of regulating the tapping of the natural African rubber tree,
Funtumia elastica, which occupied much of the Departmental energy until
the practice was eclipsed by the introduction after the First World War of
plantations of _Hevea brasiliensis_, a tree that grows better in Africa than
it does in its native Brazil. In 1903, H.N. Thompson was appointed
Conservator of Forests, in charge of the Forestry Department. He had
trained at Cooper's Hill College in London for the Indian Forest Service,
and already had 12 years working experience in Burma (18, 21).
The forestry service in Nigeria predates that in Britain - the
British Forestry Commission only coming into existence in 1919 after the
stresses of the First World War - and also predates the Forestry Bureau of
the United States of America, which was founded in 1905. In fact, the
Protectorates of Northern and Southern Nigeria had themselves only come
into existence in 1900; and the Colony and Protectorate of Nigeria was not
declared until 1914. This early awareness of the importance of forestry
owed to the British experience in India, where German foresters had been
employed to develop tropical silviculture from the European tradition of
forestry. These included such famous names as Sir Dietrich Brandis, who
was placed in charge of the entire forests of Burma in 1857, and in 1864
became the first Inspector-General of Forests in India; and as Sir William
Schlich, his subordinate, who became Professor of Forestry at Coopers'
Hill College in 1885, and in 1905 was appointed the first Professor of
Forestry in the University at Oxford. The first Governor General of
Nigeria was Frederick Lugard, the son of missionaries in India, and an
army officer who had seen service in India and Burma. He had spent a sick
leave fighting slave-traders in East Africa, and came to Nigeria
originally in 1894 as Commandant of the Niger Constabulary. At a time
when lives of Europeans could be numbered in months rather than years, he
was a survivor - from bullets, poisoned arrows and fever (19). Besides
the gift of considerable courage, he had formidable foresight. His
'Political Memoranda' published in 1919, the year of his retirement, had
an important influence on the philosophy and conduct of British imperial

policy (13). His Memorandum 13 spelled out Nigerian Government policy on forestry, and had a validity which lasted until about 1970, when the oil-boom decade transformed the economic spectrum.

Based on Indian experience, reservation of a third of the total land area for forestry was regarded as desirable for a well populated country; and a minimum target of 25 percent was proposed for Nigeria. In fact, up to the present day, only 10 percent has been reserved of which 1/5th is high forest and 4/5ths are savanna woodland with a low stocking of trees. Compare this with the forest estates in France of 25 percent, West Germany 30 percent, U.S.A. 33 percent, Japan 67 percent, and India today 24 percent. Britain has only 10 percent of forest lands and imports 90 percent of its wood requirements.

The southern parts of Nigeria had never been conquered, but ceded by treaty, so that traditional land tenure remained intact. Therefore forest reserves could only be established with the agreement of the traditional land owners. As communal land tenure was usual, ownership of the reserves remained vested in the Local Authorities (or 'Native Authorities' as they were then called). The forestry fees were paid to the local authorities, although Government retained a part of these to meet the costs of its supervisory activities in forest management. In fact most of the forestry staff were local government personnel, whose work was controlled and monitored by the Colonial Government Forestry Department. (This also had the advantage that staff could be employed who did not meet the stringent educational requirements for Government service, for which there was an insufficient supply of qualified candidates in those days). In effect the Government became the trustee responsible for management of the forest estate.

Following guidelines laid down by Thompson and Lugard, the constitution of forest reserves required elaborate procedures for consultation between Government and governed, including different departments of Government as well as local communities. The Inquiry Report included a First Schedule, which described the boundaries in precise detail, and a Second Schedule which listed the rights in the reserve admitted to local communities. Existing rights practiced in the reserve were investigated via village meetings, as survey and demarcation of the reserve proceeded; and the boundaries were marked out by cut lines or by permanent features such as rivers and streams. Rights were admitted as long as they did not appear likely to conflict with the proposed objects of management in the reserve, and included traditional activities such as hunting, fishing and gathering for food, medicines and materials for house construction and domestic use. The period from the time of proposal to the official gazettement of the reserve was not less than about 18 months, and commonly several years. Preference was given to blocks of vacant land, often land between different communities, especially where these contained valuable forest; but protection reserves for watersheds or unstable terrain were also included. In the high forest areas, forest reservation was substantially completed during the 1920's and 1930's, although there has been some reservation in the Rivers State during the 1970's and 1980's. In the savanna areas of the north, much of the reservation was made during the 1950's and 1960's (13, 21).

9.3. TAUNGYA FARMING

At the time of colonization, the population of Nigeria was only about a tenth of what it is today. The rain forests were little inhabited, and

indeed were an obstacle to penetration by Europeans from the sea to the
more populated savanna hinterland, which was only accessible via navigable
rivers. In a region where, otherwise, head porterage was the rule, a
column of porters had difficulty carrying sufficient food to see
themselves through to the other side. Hence the earliest European
explorers to Nigeria arrived from the north across the desert, not from
the sea. In 1926, a colonial forest officer, Richard St. Barbe Baker, was
posted to Sapoba to open the forestry station there, which lay in the
middle of an extensive forest area. In order to feed his forestry workers
and their families, he introduced 'Taungya' farming, which he had
experienced in Kenya where he had served previously. It is significant
that 'Taungya' is a Burmese word meaning 'Hill farming', and was an
adaptation of the formerly widespread practice of shifting cultivation to
the establishment of tree crops (5,8).

Natural tropical high forest consists of a mixture of a wide range of
species and size classes, of which on average only about 5 trees per
hectare were then marketable. The system of planting stump sites had been
found to be unsuccessful due to the impossibility of caring for the
planted seedlings (4,21). It is difficult to find methods of regenerating
natural forest which can be afforded from a part of the government revenue
from exploitation. In the rain forest zone it could be said that the
forest lives off the products of its own decay, and a major proportion of
site fertility resides within the plant biomass itself (15). Traditional
peasant farming makes use of this by felling and burning a patch of
forest, which is cultivated for 2 or 3 years, in order to utilize the
residual fertility; and then the natural vegetation is allowed to
reestablish itself. It may require 50 years before the soil has fully
recovered its fertility, and more than 100 years before the forest
approaches its former condition. Nevertheless, by planting trees of
economic species among the arable crops, a commercially valuable tree crop
can be established - more valuable than the natural forest which it
replaces, because almost every tree becomes utilizable. This is what
Barbe Baker proposed should be done.

In 1927 the Nigerian Government appointed 2 silviculturists to the
Forestry Department: W. D. MacGregor centered at Olokemeji, and J. D.
Kennedy working at Sapoba. These officers had been selected from among
the staff of the Department, and were then sent on a tour of silvicultural
activities in India and Burma as part of their training. They gathered a
substantial body of information on the composition of natural forest, and
Kennedy published the "Forest Flora of Southern Nigeria" in 1936 (9,14).
By means of experimental trials they identified species suitable for
timber crop plantations, and also initiated research into methods of
natural regeneration, and of forest enrichment by means of line planting.
Kennedy established 8 hectares of experimental plantings each year, using
the Taungya system. In the drier high forest reserves, the practice of
Taungya was found to be essential for successful establishment of Teak,
Tectona grandis, which had been introduced from Burma. Their work
continued until brought to a halt in 1935 by economic recession; but it
laid the foundations for all subsequent silvicultural management in the
Nigerian high forests.

In 1936 Taungya schemes were initiated on a management scale in the
Benin forests, and forestry settlements were established at various
localities within forest reserves, and particularly within the Sapoba
Forest Reserve. For each forest community, farms were demarcated each
year for arable cultivation. In this region, the creation of forest

reserves, the growth of population, and the expansion of cash cropping (usually of tree crops such as cocoa or rubber) have caused a shortage of land for raising food crops - and in a peasant society each community is more or less responsible for feeding itself. The peasant farmers were prepared to clearfell and burn the forest without payment in order to grow food crops for themselves. They were content to do this because they obtained fertile virgin land to cultivate, but were able to maintain perennial crops such as citrus, cocoa, cola or rubber on their own lands (and these are much less detrimental to the soil). Traditionally their arable farmlands are some distance from the community, in order to be inaccessible to freeranging village livestock. However, farmers are not normally prepared to walk more than 5 km to their farms. The Forestry Department acts as 'Chief farmer', surveying and dividing up the farms, and ensuring that all farmers brush and clear the land and plant their crops at the correct times. This is important, as all the farms must be ready for burning at once, because a premature burn or a badly felled plot leaves a considerable amount of debris that must be cut, stacked and burned in bonfires, greatly adding to the farmer's labor. A good burn reduces the incidence of pests and weeds, and the contiguity of the farms reduces damage from forest vermin. Moreover the Forest Department may sometimes assist with felling larger trees. In return, among the peasants' arable crops, the Department is able to plant trees, which also benefit from the cultivation given to the soil by the farmers. Formerly the farmers were paid a bonus for successful establishment of a tree crop; and the size of the plot granted in a subsequent year, or indeed whether any plot is granted at all, may be conditional on tree survival within a farmer's plot during the previous year. It appears that the better farmers also obtain a better survival and growth of the tree crop. Thus there should be mutual rewards: the foresters replace the 'bush fallow' with a valuable timber crop, and the farmers are able to benefit from the residual fertility of the original forest.

In the high forest zone, peasant farmers have traditionally practiced mixed cropping - that is combinations of melon/yams/maize/cassava into which other crops such as okra, amaranthus, tomatoes, peppers, sugar cane and other vegetables are interplanted (Table 1). The peasant needs to get maximum production from minimum area, both to feed his family and to sell, because of the limited area that he can clear and cultivate from his own resources. Mixed cropping is usual, and has been shown to increase overall yields from unfertilized fields. The farmer and his family do most of the work themselves, relying on axe, hatchet and short-handled hoe for their work. They can clear between 1/4 ha and 1 ha annually, and also maintain fields from the previous year or two years. The farmer under-brushes the forest, fells the trees, burns the debris and digs the mounds, and the root crops belong to him. His wife helps with weeding, harvesting and marketing the produce; and the vegetables and peppers belong to her. They are able to sell between a quarter and three-quarters of their total production. Farmers have traditionally always retained fruit trees on the farm. These include _Chrysophyllum albidum_, _Cola acuminata_, _Dacryodes edulis_, _Dialium guineense_, _Irvingia gabonensis_, _Pentaclethra macrophylla_, _Treculia africana_, and the oil palm _Elaeis guineensis_. The Forestry Department fits well into this traditional system, and in return for furnishing the land and organizing the farmers, the Department has its own benefit which is the establishment of a tree crop. The last arable crop normally grown is cassava; and after 2 or 3 years of farming, cultivation is abandoned and a bush fallow is allowed to

Table 1.   The farming cycle (11). The tree crop is planted in Year 1,
           during May and June.

	Ogun & Ondo States	Bendel State	Cross River State
Annual Rainfall:	1350 - 1600 mm	1500 - 2000 mm	1700 - 2600 mm
Geology:	Basement-complex	Sedimentary sand	Basement-complex
Farming agency:	Peasant farmer	Peasant farmer	Forestry Dept.

Year 0

October	-	-	Demarcate farms
November	Demarcate farms	-	Brushing
December	Brushing	Demarcate farms	Felling

Year 1

January	Brushing	Brushing	Burning & packing
February	Felling	Brushing	Burning & packing
March	Burning & Packing	Felling; Burning and packing	Sow maize (1x1 m)
April	Sow maize, Egusi melon, Okra; Dig yam mounds	Sow maize, melon, Okra, Amaranthus, & peppers	Sow maize
May	Plant yams when maize germinated	Dig mounds & plant yams	
June	Harvest maize	-	Harvest maize
July	Harvest melon	Harvest maize & melon	Plant cassava
August	Plant cassava	Harvest maize	Plant cassava
September	Resow maize	-	Resow maize
October	-	-	-
November	Harvest 2nd crop maize	Harvest yams Harvest yams	Harvest 2nd crop maize.
December	Harvest yams	Harvest yams	-

Year 2

January	-	Plant cassava	-
April-August	Harvest cassava	-	-
July-December	-	-	Harvest cassava

Year 3

January-March	-	Harvest cassava	-

grow, mainly from coppice shoots and root suckers from surviving tree
stumps (although in Taungya this is replaced by the plantation tree crop).
At Sapoba, the soils are deep unconsolidated sands, and lose fertility
rapidly by leaching, due also to depletion of soil humus following forest
removal, which occurs rapidly in a warm moist climate. Farmers may be
reluctant to return to the same areas to cultivate for at least a decade,
although secondary forest is less arduous to clear than primary forest.
After resting 6 or 7 years the soil may recover about half the fertility
that it had on initial clearing.

In the Benin forests, which include Sapoba, the Forestry Department
planted Nauclea diderrichii as a nurse crop to various mahogany species,
such as Entandrophragma cylindricum, Khaya ivorensis and Lovoa
trichilioides, at a planting spacing of 12 feet (3.5 metres, Table 2).
These are planted during May to July in the first year of farming, and
close canopy after about 6 years. The planting of perennial crops was not
allowed, except possibly of plantains at the edges of their farms; and the
planting of cassava at the beginning of the farming cycle was discouraged
- in order to avoid smothering the tree crop. Initially, for the same
reason, the Forestry Department forbad the planting of cassava, but later
the planting of erect varieties was allowed, and cassava is now a normal
component of Taungya farms. It has proved impossible to grow mahoganies
as a pure crop, because of their susceptibility to attack by the insect
shoot-borer, Hypsipyla, and because they have narrow crowns and tend not

Table 2. Planting spacings used for Taungya tree crops (modified from
11).

Tree crop	Planting stock	Planting spacing (m)	Trees per hectare	Locality (State)
Gmelina arborea	stumps & polipots (7.5x12cm)	3x3 3x2 2.5x2.5	1076 1667 1682	Ogun & Ondo (pulp projects) Cross River
Nauclea diderrichii, Meliaceae (mahoganies) in 5/1 mixtures	stumps	3.5x3.5	747	Bendel
Tectona grandis	stumps & polipots (7.5x12cm)	2.5x2.5	1682	Ogun & Ondo
Terminalia ivorensis	stumps	5x5 5.5x5.5	420 332	Ogun & Ondo Bendel

Note: Polipots of 3x5 ins flat (7.5x12 cm) are also suitable for Nauclea
and Terminalia, and give better survival and growth than stump plants.

to control weed growth by canopy closure. However, establishment and
growth have been satisfactory where they are mixed with Nauclea as 5 nurse
trees to each mahogany. The Nauclea can be coppiced for transmission
poles after 15 or 20 years, leaving the mahoganies to grow on a timber
rotation of about 70 years. A fifth of the area of the farms was planted
with pure crops of Terminalia ivorensis at a spacing of 16 feet or 18 feet
(5 m or 5.5 m, Table 2), which require a rotation of 35 years to reach
timber size (20). Other suitable species on these soils include Antiaris
africana, Aucoumea klaineana, Cedrela odorata, Terminalia superba and
Triplochiton scleroxylon. The system appeared so successful that, during
the 1940's, it was proposed to convert all the Benin forests progressively
by the Taungya method (10), although the tempo would have been constrained
by the inflow of funds derived from fees for forest exploitation. These
were also important as a source of revenue to the local government
authorities, to finance their administrative and welfare activities in the
community.

9.4. PREFERENCE FOR THE TROPICAL SHELTERWOOD SYSTEM

In 1948 the first integrated sawmill and plymill industry began
production at Sapele, 30 km from Sapoba, in response to the insatiable
demand for timber generated by the need to reconstruct Europe after the
Second World War (23). Forest exploitation was stepped up, and almost all
the high forests came under working plans, operating on a felling cycle of
100 years. It was thought desirable to find a system of regeneration that
could keep up with the progress of forest exploitation; it was clear that
the use of Taungya, which had been successful on a pilot scale, was beyond
the resources of the Forestry Department to the extent that had become
needed; nor (in those pre oil boom times) could Government have found the
necessary funds from fees paid for timber removals or from other
Government revenues. Moreover there had to be sufficient peasants who
were prepared to co-operate in such schemes, which was not everywhere the
case. It was necessary to use a method that was cheap enough to be
financed from a part of the earnings from fees for forest exploitation,
which because of low removals from extensive areas of forest, were very
modest.
Consequently, in 1950, the Forestry Department prohibited any
augumentation of Taungya operations - although existing schemes were
allowed to continue - and the Department turned to natural regeneration as
the means to restore exploited forest. Based on experimental trials by
Kennedy and others, and encouraged by experience in Malaya (from whence
several officers had been evacuated to Nigeria during the war years), the
Department applied the Tropical Shelterwood System (T.S.S.) of
regeneration. This consisted of a series of preexploitation treatments
during a period of about 5 years, with climber cuttings and cuttings of
non-economic species in the thicket, and poisoning unwanted trees with
solutions of sodium arsenite poured into frill girdles cut into the bark.
In the decade of the 1950's, about 2000 sq. km of forest were treated in
this way. It was intended to promote survival and growth of both
seedlings and trees of valuable species. A minimum of 100 well grown
seedlings per hectare, and final crop of 25 fully grown trees of economic
species, was regarded as satisfactory regeneration. However in 1952, the
eminent tropical silviculturist, A. Aubreville, wrote to the Chief
Conservator of Forests concerning T.S.S.: "I am set in my skepticism
regarding the application of these methods except where the pre-existing

regeneration is abundant.... It is tempting for a forester for, after having interfered with the forest in order to favor natural regeneration, he can continue to maintain hope of success for a pretty long time, and if it has no result it is not obvious in the forest... If it is a matter of ... a programme of work involving a considerable expense one must be sure of success and profit." Whereas, in Malaya, during a regeneration survey minimally a third of milliacre quadrats were expected to contain established regeneration, this was about the best that could be obtained in Nigeria. Moreover, there was a tendency to poison species of trees which subsequently became exploitable, and also to cut the regeneration of economic species along with the remainder of the thicket (12). It also seems that the potential economic returns on expenditure from the system were too low to be attractive to Government; information which may have been suppressed by the Forestry Department at the time, in order that all investment in silviculture should not be discouraged.

9.5. RESUMPTION OF TAUNGYA

In 1954, three Regional Governments, Western, Eastern and Northern, were created under the aegis of a Federal Government. The management of the forest estates became a function of each Regional government, and only forestry education and research remained Federal responsibilities. After national independence in 1960, in the Western and Eastern Regions the British forest officers were replaced by Nigerian staff within three years - although in the Northern and Federal services the former British personnel continued for a period of about 15 years after Independence. However cogent the silvicultural arguments against employment of T.S.S. in Nigeria may have been, these were not the direct reasons for its abandonment. In 1963 the Mid-western Region was split off from the Western Region, and Nigerian officers were redeployed to their Region of origin. These staff disruptions interfered with the management capabilities of the forestry services. Though the written forest policy remained the same, formerly more junior officers were able to make differing priorities felt. At this same time the forest estate came under pressure from other forms of land use, particularly cocoa, oil palm and rubber plantations, as well as from peasant arable cultivation. The Forest Reserves could not continue to be used at such low intensities of utilization, and something more obvious than natural regeneration had to be employed if the land were to remain in forestry use. There was, therefore, a shift from T.S.S. back to artificial regeneration, particularly by means of Taungya. In the 1960's the planting of Teak was resumed in the Gambari and Olokemeji Forest Reserves, using Taungya methods. Planting of Teak in these reserves had begun in 1915, and during the 1920's and 1930's extensive areas had been established. Later, because of the incidence of root and butt rot, from the fungus _Rigidoporous lignosus_, these plantations had been converted to coppicing rotations to produce firewood; but in the 1960's the coppice was restored to a timber rotation because of the high value of teak as timber (2). In the Western Region, the impending need for Nigeria to produce its own cellulose pulp for paper manufacture was perceived, and in 1966 A.K. Jaiyesimi of the Western Region Forestry Department began to establish extensive plantations of _Gmelina arborea_ at Ajebandele in Omo Forest Reserve (3). These various plantations served to form a belt along roads and around settlements, making land available for peasant arable crops in conjunction with tree crop establishment, but ultimately barring access to

the forest land behind. The expense of these plantations was justified by
the protection that they gave to much larger areas of natural forest.

Also in the 1960's, in the Cross River North Forest Reserve near Ikom
in the south-east part of the country, an attempt was made to develop
cocoa Taungya - that is to establish cocoa together with a timber crop.
The work was undertaken by the Eastern Region Forestry Department
in collaboration with the Eastern Region Development Corporation. This
proved a mistake, and in effect alienated the land to the Development
Corporation, who began to fell the timber trees before they could reach
merchantable sizes in order to favor the cocoa crop. In Nigeria, the area
of high forest reserve is only 2 percent of the total land area, which is
inadequate to meet the existing needs of the nation, let alone those of a
developed economy. It is clear that, although combinations of timber
crops with other perennial crops such as cocoa, coffee or bananas are
practicable, they should not be used within forest reserves where the main
object of management must be sustained production of wood.

The area allotted to each Taungya farmer is generally about 0.5 ha
per year, varying between 0.25 ha and 1 ha. About 60 percent of farmers
originate within 8 km of the Taungya centre. A half of the farmers are
more than 45 years old and have been involved with Taungya for more than 8
years - and some for over 30 years. The family size averages 1.5 wives,
2.4 children under 8 years old, 4.0 children above 6 years old, and with
6.0 other relatives in the household. The children work only about 4
hours per week on the farm, and this may have been further reduced with
the advent of universal primary education in 1976. Three quarters of the
farmers hired labor to assist with their work. In 1975 there were about
25,000 Taungya farmers and about another 20,000 people obtained casual
work on the farms. Average yields were: maize grain 1 tonne/ha, yams 5
tonnes/ha and cassava 10 tonnes/ha. The families themselves ate between a
quarter and three quarters of their total production, and usually more
than half. Over half of the cash earnings were spent on hired labor.
However, most farmers had additional sources of income, such as from other
farms, or from crops of cocoa, rubber or oil palm on their own lands; or
from paid employment such as with the Forestry Department. In 1975 the
total area of peasant Taungya in southern Nigeria was about 20,000 ha.
The estimated total food output from these farms was 11,500 tonnes of
maize, 143,500 tonnes of cassava and 76,500 tonnes of yams, then worth
altogether about $1 million (6).

It is also possible to practice Taungya in savanna forest reserves,
although this is less usual. For example, plantations of Neem,
Azadirachta indica, can be established with crops of groundnuts or
cowpeas. This is also possible for tropical pines such as _Pinus caribaea_,
or _Pinus oocarpa_ in the drier more northerly Guinea savanna. However in
the case of _Eucalyptus_ species there may be more difficulty, because they
are sensitive to termites. Protection can be given by a dose of the
long-lasting soil insecticide 'Dieldrin', incorporated in the seedling
potting medium. However, after planting the seedlings out in the field,
cultivation can push soil over the top of the potting medium which
protects the root collar, forming a bridge that allows termites to attack
and kill the young tree. In the drier savanna areas the tree crop may
also be susceptible to competition for water by the arable crop, or by
woody growth surviving from the natural vegetation. In these
circumstances it may be better to keep the ground between the trees well
cultivated and bare, until the tree crop closes canopy. This also reduces

the risk of fire damage to the tree plantation by reducing the availability of combustible material during the dry season.

9.5.1. 'Direct Taungya'

Taungya activities in the high forest zone were intensified during the Nigerian Civil War which lasted from 1967 to 1970. In the southwest of the country these Taungyas were an important source of food to the towns at a time of economic stress, which restricted food importation; and about 2,500 ha of new land were brought into cultivation each year by this method. In the south-east the 'Biafran' enclave was under siege, and it was to the Forestry Department that the rebel authorities turned to organize food production. The effects of these lessons were not lost after the Civil War, and in the Southeast State (now Cross River State) the Forestry Department became directly engaged in both food crop production and establishment of timber crop plantations, particularly of Gmelina and mainly by the 'Direct Taungya' system for which E.U. Okon was the chief architect. Funds supplied for the relief of Nigerian refugees from the Macias regime in Fernando Po (now Equatorial Guinea) were used to finance these operations as well as to provide useful work for the refugees themselves. The main crops were maize and cassava, the cassava being converted to 'gari' (i.e., cassava flour) using the traditional hand methods. These crops are well suited to extensive operations, and at each project centre it was possible to plant about 500 ha annually. The Forestry Department itself had become both farmer and forester, employing workers to carry out the agricultural as well as the silvicultural operations, but selling the produce itself in exchange for daily wages to the workers. By the mid 1970's, in the south of Nigeria, the State Forestry Departments, sometimes with the assistance of Federal grants, were reforesting a total of about 15,000 ha each year, of which about 10,000 ha were by Taungya - both direct taungya and with the assistance of independent peasant farmers (16) (Table 3).
    In 1980, it became necessary to supplement the earlier plantings of Gmelina by the former Western Region Government in both the Omo and Oluwa Forest Reserves, in order to feed the pulp and paper mills under

Table 3. Progress of reforestation by planting (ha) in the main forestry States of southern Nigeria. About two-thirds of the area was regenerated by Taungya (either peasant Taungya or 'Direct' Taungya).

States	1940	1950	1960	1965	1970	1975	1980
Ogun			1 000	1 490	5 210	13 170	24 300
Ondo	700	750	-	1 316	3 678	14 152	19 300
Oyo			123	608	3 138	6 056	13 600
Bendel	430	3 790	7 177	11 900	23 350	41 711	44 100
Cross River			1 300	4 700	4 700	13 580	17 970
Total			9 600	20 014	40 076	88 669	119 270

construction at Iwopin. The World Bank joined with the Federal Government and the governments of Ogun and Ondo States, with the intention of planting about 3000 ha annually at each centre, with a view to establishing eventually a total plantation area of about 40,000 ha - although there has been difficulty in exceeding 2000 ha annually in either of the project sites. Consequent on the oil boom, the rising standards of living in Nigeria, and the attempt to 'democratize' education, people and especially young people have become unwilling to engage in peasant farming. It has therefore become necessary to mechanize both the land clearing and arable cropping operations in order to obtain timely completion of the work. Felling has been by bulldozers fitted with shear-blades, cutting trunks and woody growth at ground level while leaving roots in the soil. After burning in situ, the debris are pushed into windrows in order to facilitate planting operations. The maize is drilled in during April, and during the following 2 or 3 months the Gmelina tree crop is planted amongst this at a spacing of 2 metres in rows and 3 metres between rows. This may be followed by a crop of cassava. More recently rice has been introduced, and also a catch crop of cowpeas following the maize. There is no second crop of maize during the year, as yields have proved too low. In 1981 arable cropping was carried out over an area of 100 ha at each project site, but this has increased progressively to 600 ha for each project in 1985. Yields of maize do not commonly exceed 1 tonne, and yields of cassava average about 8 tonnes. These are comparable with the yields obtained by peasant farmers. However, the projects apply preemergent sprays to control weed growth, and also give a dose of fertilizers. The maize can be sold for about $600 per tonne, compared with a world price of about $100, and the cassava for about $200 per tonne. The Gmelina closes canopy within 2 years. Ball and Umeh (7) estimate internal rates of return as follows:

		Gmelina	Teak
Traditional Taungya	1975	11.3%	5.5%
Direct Taungya	1975	11.9	5.6
Direct planting	1975	5.8	4.0
Direct Taungya	1980	18.0	6.6
Rotation		15 years	60 years

The improvement in the rate of return between 1975 and 1980 is due to the increasingly advantageous price of food in Nigeria in relation to costs, and this will have continued to improve. Moreover the price of maize was only $200 per tonne in 1975, compared with $400 in 1980 and over $600 today.

9.5.2. Decay of Taungya

In 1976, E. U. Okon left Government service, and by 1978 direct Taungya operations in the Cross River State were largely discontinued. A contributory reason may have been the ending of Federal relief funds for the refugees. It is possible that the money realized from the sale of farm produce did not meet the cost of the agricultural operations, and made the State Government unwilling to continue with the work. My personal view is that the World Bank schemes at Omo and Oluwa will also decay once Bank support has been withdrawn, although the plantations may continue to supply pulpwood to the pulp mills when these come into operation. The reasons are the inadequate funding of the State forestry services, in a situation where there are more pressing urgencies for State funds than staving off the exhaustion of the nation's timber supplies, forecast by FAO to occur by the year 1995 in the forests of the southwest and centre of the country (1). Another result is that the forestry services have tended to become demoralized and inefficient. As regards traditional Taungya, political pressures have caused land to continue to be released for arable cropping by peasants, even though there were insufficient funds to ensure the planting of a tree crop. The consequence has been to destroy extensive areas of natural high forest within forest reserves, leaving seas of the shrubby weed Eupatorium (Chromolaena odorata) or tree weeds such as Musanga cecropioides, Macaranga barteri or Trema orientalis, or tangles of lianes and climbers. Sometimes substantial areas of young plantations are burned due to the activities of disenchanted villagers, or unpaid laborers. It is doubtful if traditional peasant Taungya could now be operated successfully on more than a limited scale.

9.6. THE ROLE OF PEASANTS

The attitude of peasants to forest reservation has been various. At the time of reservation of the high forests, over fifty years ago, the population was about a quarter of what it is today, and cash cropping was much less extensive. Bush was plentiful, and the people could not contemplate a possible need for all the land for their own use. Customarily, undeveloped land was communally owned, and, as after reservation it remained under the aegis of the Local Authority presided over by their own chiefs, the peasants might perceive no considerable change in its tenure. They were persuaded by the Forestry Department that forests would continue to be needed, not only for timber production but also for the materials that were essential for their own livelihood; and that the forests would disappear piecemeal if they were not protected. Moreover, reservation appeared to give legal reinforcement to the ownership of the land by the local communities, and could help to protect it from expropriation by strangers. The people were in awe of their chiefs, and of the Government, and as long as their existing rights were preserved they were not usually disposed to seriously contest the establishment of a forest reserve. However the Oba (King) of Benin, who was the traditional trustee of the land for his people, would not lend his authority to the establishment of forest reserves unless it was agreed that access by peasants for arable cropping would be allowed within forest reserves by means of Taungya farming (4).

Along the west coast of Africa, there have long been commercial links with America and Europe: firstly via the slave trade, and secondly via the commerce in palm oil - which washed, fed, lighted and lubricated the industrial revolution in Europe. There was also trade with tribes in the hinterland, to whom were sent cola nuts, dried fish, gold, ivory and manufactured goods from Europe; and from whom were received in exchange beans, corn, copper, silver, salt, cloth and livestock. Colonization, and the enforced peace which it brought, encouraged growth of the export trade, of which the peasant rapidly took advantage. Cocoa and cola (Cola nitida) were introduced on their own initiative, and for example the annual production of cocoa increased 20 times during each succeeding decade between 1885 and 1915, 5 times between 1915 and 1925, and twice between 1925 and 1935. Hevea rubber was made available to peasant farmers during the 1920's, to replace the natural African (Funtumia) rubber, with

a similar experience to that for cocoa; it could grow on soils unsuitable
for cocoa, and peasant farmers adapted well to cultivating rubber (17).
Oil palm has remained mainly a naturally grown tree. An effect of
planting up peasant holdings with perennial crops was to alienate land
from communal to individual ownership, and to instigate an evolution
towards freehold land tenure. But this received a set-back from the Land
Use Decree of 1978, which has restrained this trend, although the Decree
is unpopular in rural areas. In Nigeria, the policy of Government was
formerly to encourage peasant holdings, and to discourage foreign-owned
plantations (13), although recently the Government has appeared to
encourage the establishment of commercial plantations in order to hasten
industrial development.

The increasing area of perennial cash crops has also led to a
Although Government has always ostensibly encouraged cash cropping by
peasants, the exploitive tendencies of Government-controlled Marketing
Boards, combined sometimes with stagnating or falling world prices for the
produce, have resulted in a falling production of export crops. This has
also occurred for food crops, and for other cash crops for the home
market, where Government has attempted to regulate the price. The fact is
that peasants are not easily exploitable, because they are still basically
subsistence producers. If they feel that they are not getting proper
value, they can (and do) largely withdraw from cash crop production, and
still live from their own resources. The burgeoning of Government
employment, and of urban growth, has meant that education is a more
certain route to wealth and status than peasant farming; and peasants are
investing in the education of their children, rather than in farming.
Peasant farming has become the last resort of the unlucky, the uneducated
or the unaspiring.

The increasing area of perennial cash crops has also led to a
shortage of land for arable cropping. In forest areas, peasants do not
customarily manure their fields, but allow them to recover under a bush
fallow. However, on short fallowing cycles, the land available to them
for raising food becomes impoverished under repeated cultivation. The
peasants have therefore been willing to continue their former practice of
shifting cultivation on virgin soils within forest reserves, organized as
Taungya schemes by the forestry department. Forestry employees have also
been happy to join in these schemes, and to obtain access to land in order
to feed their families. The relatively low standard of living of
peasants, and the domination of village life by older people, encourages a
drift of young people to the towns, and to 'aging' of the farming
population, the average age of farmers now being between forty and fifty
years. Young people prefer wage earning or salaried employment, in order
to have money in their pockets. This desire can be partly met in rural
areas by plantation agriculture (including forestry). What had existed of
this, has been in the hands of Government, or of international companies,
who have not been exploitive of their employees. However, there is the
danger that if it is abused the system can attract the odium that it has
in other parts of the world.

During the 1960's, particularly in Bendel State, Taungya farmers were
obliged to cultivate larger areas than they may have wished or could
conveniently manage, because of the desire of the Forestry Department to
increase the area of forestry plantation. However, during the past two
decades the price of rubber has slumped, and there has been difficulty
hiring labor to tap the trees. In the Ondo and Oyo States peasant cocoa
plantations have been becoming unproductive, with trees dying, due to
overmaturity. This was not anticipated by the farmers, and the result is

a drastic fall in their earnings. In these circumstances the farmer
cannot afford to hire labor to replant and to wait several years until
young trees come into fruit. Moreover, because farmers tend to be no
longer young, they are not capable of doing the work themselves.
Consequently, peasant land holders turn to food crops as cash crops, which
has renewed the pressure on forest land for Taungya. However, State
forestry departments have become unwilling or unable to pay a bonus to
Taungya farmers for good performance. Moreover, the farmers themselves
may be expected to plant the forestry trees without payment. Sometimes
the Department fails to supply tree seedlings to the farmers, whether due
to lack of sufficient nursery stock, or to lack of transport, or to
administrative difficulties. In consequence, Taungya farmers have become
uninterested in the survival and growth of the tree crop, and may avoid
planting the trees, or damage or destroy them afterwards in order to favor
their own crops or to protract the period of cultivation. This may go
unchecked by the forestry department, who are 'politically' weaker than
the farmers. A not uncommon result is extensive clearings within forest
reserves for arable cultivation, without the subsequent establishment of a
tree crop. During 1984, because of these difficulties, the Bendel State
Government forbade the clearing of more forest for Taungya farming, but is
under considerable pressure to accede to the demands of peasant farmers
for more land to farm within forest reserves.

Forestry plantations, whether established by Taungya or by direct
planting, change the land use within forest reserves. The forest is
substantially destroyed as a source of the materials needed by a rural
community, and is also changed as a habitat for the wild animals which are
an important source of animal protein. Furthermore, the collection and
sale of 'minor' forest produce from natural forest provides an important
income to rural communities throughout the year - from bush meat, snails,
'native sponge' (a liane), wrapping leaves, chew sticks, peppers, spices,
fruits, medicines, etc. However, these items may be exploited by
strangers, who pay a nominal fee to the local authority for the
permission, or may remove them without a by-your-leave from anybody.
Today, the accessible forests are being over-exploited for both timber and
minor produce, to an extent that is destroying the resource base.

The Forestry Department is not particularly popular in rural areas,
being perceived rather in the guise of 'bush policeman'. Forest guards,
although locally recruited, may take advantage of the ignorance of
peasants to bully them, or to extort 'fees' for what ought to be free, or
for what was formerly free. This abuse of power owed particularly to
legislation that established 'protected trees' - with the motive of
preserving species which yielded valuable timber or fruits. Forestry
Department permission may be needed to fell them, for which a fee may be
payable to the local authority, who regarded this as a useful source of
revenue. Forest officers had magisterial powers, and were authorized to
seize property or to exact summary fines on persons infringing forestry
regulations, although their findings could be challenged in the courts.
With increasing land shortage, peasants have begun to resent alienation of
their land to forestry, and this has been exacerbated by the control of
forest reserves being usurped by State Governments since the Land Use
Decree. Peasants and chiefs may now have little say in the management of
the reserves. Timber concessions may be handed out to strangers (although
possibly from the same State) without local communities being consulted,
or deriving any material benefit. There are instances of reserved land
being allocated for establishment of commercial plantations of cocoa, oil

palm or rubber, or for some Government agricultural scheme, without
acquiesence from the former communal land owners.

9.7. CONCLUSIONS

Taungya farming is essentially a system of shifting cultivation,
organized by the Forestry Department within forest reserves, as a method
of replacing natural forest with timber crop plantations. The growth of
woody biomass of natural forest is low, estimated as an annual increment
of 2 cu metres/ha compared with 25 cu metres/ha for _Gmelina_ plantations,
and the objective is to replace it with more intensive forestry use at
minimal cost. However, the method relies on having a supply of people
willing to do this work, and depends on the existence of a peasant
population in a substantially subsistence economy at a low standard of
living. In Nigeria today this is not a situation which people suffer
gladly. The foresters, of course, welcome the assistance of the peasants,
who clear the forest at nil or minimal charge, and whose cultivation of
the soil also benefits the tree crops. This is particularly advantageous
in a developing country, where funds available for forest regeneration are
small. For their part, the peasants gain access to virgin soils for
arable cropping, when their own lands are becoming exhausted, or which
they had converted to more profitable tree crops, such as citrus, cocoa,
cola, oil palm or rubber.

Outside forest reserves, peasants may use the method for the
establishment of their own tree crops. However they are not greatly
interested in a crop which may take ten, twenty, thirty, fifty or seventy
years to mature; the economic advantages of timber production are
irrelevant to them. Moreover, under the existing forest laws, payment of
fees to Government is due on listed tree species, whether these are
naturally growing or have been planted by the owner; and this also serves
to discourage commercial tree planting. The obvious conclusion is that
the Forestry Departments must be prepared to do the arable cropping for
themselves, both as the need for natural forest conversion intensifies and
as the general standard of living rises. The Forestry Departments have
done this successfully on a pilot scale, but are proving incapable of
achieving the large-scale reforestation now needed, if the country is to
produce its own timber requirements. The reasons are the unsuitability of
Government bureaucracy for what is essentially commercial enterprise, the
difficulties of running large bureaucracies in developing countries, and
because State governments have more pressing needs on which to spend
limited financial resources than for growing trees.

Nevertheless, in Nigeria, the Taungya system is well practiced and
well understood. It is possible to plan a project and carry it out,
substantially according to the methods and costs worked out in the
original feasibility study. This is more than can be said for most
agricultural projects in the lowland moist tropics. The system is robust,
as the present cost of operating manually is very similar to, or less
than, that of using highly mechanized methods. This means that external
inputs can be relatively small, and vulnerability to the vagaries of the
economy is reduced. Moreover, hand labor may produce somewhat higher
yields, both of the arable and tree crops, than the mechanical methods.
Peasant Taungya can continue where people are still prepared to do it, and
is still perhaps the cheapest method of forest conversion. However,
plantation forestry offers wage employment, which is often preferred by
younger people; and, moreover, forestry is a relatively high employer of

labor per capital input. Nevertheless, because of their transient nature
as currently used, mechanical methods do not appear to be unduly damaging
to the soil, as may be the case for permanent arable cultivation in the
lowland humid tropics. The Taungya system made a substantial contribution
to the feeding of the towns during the Civil War, and could do so again
during the present economic emergency, when the country is attempting to
urbanize and industrialize, and to progress from peasant to commercial
farming. It ought to be possible to make development funds available
through the banking system, with management companies doing the actual
forest operations, monitored by the government forestry services. Once
such companies exist they would also be available for planting
non-reserved lands in the possession of communities or of individuals.

REFERENCES

1.   Anon. 1979. Forestry development - Nigeria - project findings and
     recommendations. FAO, Rome. 88 pp.
2.   Adekunle, A.O. 1964. A conversion system for the management of
     coppiced Teak plantations. In: Proc 1st Nigerian For Conf, pp 9-11.
     Kaduna, Nigeria.
3.   Alade, G.A. 1972. Gmelina pulp/paper project, Western State of
     Nigeria. In: Proc For Assoc Nigeria Conf (1970), pp 273-277. Ibadan,
     Nigeria.
4.   Allison, P.A. 1973. A historical sketch of the forests of Benin. Bull
     Nigerian For Dep, Ibadan, No 33. 30 pp.
5.   Allison, P.A. 1956. Working plan for the Kennedy-Ona plantation (B C
     29 unlicensed) for the period 1/4/56 to 31/3/66. Fed Dep For, Ibadan,
     Nigeria.
6.   Ball, J.B. 1977. Taungya in southern Nigeria. Fed Dep For, Ibadan,
     Nigeria. 35 pp.
7.   Ball, J.B. and L.I. Umeh. 1981. Development trends in taungya systems
     in the moist lowland forest zone of Nigeria between 1975 and 1980.
     Fed Dep For, Ibadan, Nigeria. 25 pp.
8.   Baker, R. St. Barbe (pers comm); Ibid. 1938. Silvicultural
     experiments at Sapoba, Nigeria. Empire For J 7:203-208.
9.   Kennedy, J.D. 1936. Forest flora of southern Nigeria. Govt Printer,
     Lagos, Nigeria. 276 pp.
10. Lamb, A.F.A. (pers comm).
11. Lowe, R.G. 1978. Farm forestry in Nigeria. In: Proc Ann Conf For
     Assoc (1975). Calabar, Nigeria. 12 pp.
12. Lowe, R.G. 1978. Experience with the tropical shelterwood system of
     regeneration in natural forest in Nigeria. For Ecol Management
     1:193-212.
13. Lugard, F.D. 1919. Political memoranda. Waterlow, London. (Reprinted
     by Cass, 1970).
14. MacGregor, W.D. 1934. Silviculture of the mixed deciduous forests of
     Nigeria, with special reference to the south-western provinces.
     Oxford For Memoirs No 18. 108 pp.
15. Nye, P.H. and D.J. Greenland. 1960. The soil under shifting
     cultivation. Tech Comm No 51, Commonw Agric Bureau. 156 pp.
16. Okon, E.U. 1973. Departmental Taungya system as practised in
     South-eastern State of Nigeria. In: Proc Ann Conf For Assoc. Enugu,
     Nigeria. 11 pp.
17. Opeke, L.K. 1982. Tropical tree crops. John Wiley and Sons, NY. 312
     pp.

18. Oseni, A.M. 1971. Outline of the history of forestry in Nigeria. Bull Nigerian For Dep, Ibadan, No 31. 5 pp.
19. Perham, M.F. 1956. Vol. 1, Lugard - The years of adventure 1858-1898; 1960. Vol. 2, Lugard - The years of authority 1898-1945. Collins, London.
20. Redhead, J.F. 1960. Taungya planting. Nigerian For Inf Bull (n.s.) 5:13-16.
21. Rosevear, D.R. 1952. Nigeria Handbook, Cap. XVI, pp 174-205. Govt Printer, Lagos, Nigeria.
22. Thompson, H.N. 1948. History of the Nigerian Forest Department. Farm For 9:11-19.
23. United Africa Company Ltd. 1952. Statistical and Economic Review No 10. 66 pp.

10. MANAGEMENT OF NATIVE PALM FORESTS: A COMPARISON OF CASE STUDIES IN INDONESIA AND BRAZIL

Anthony B. Anderson

Associate Researcher, Departamento de Botanica, Museu Paraense Emilio Goeldi, Caixa Postal 399, 66.000 - Belem, Para - Brazil

10.1 ABSTRACT

Native forests dominated by palms are widespread in the tropics. These forests occur in coastal and freshwater swamps and form spontaneously on secondary, upland sites. Palm forests provide people with a diverse array of market and subsistence products and support a variety of other economic activities.

This paper examines in detail two forest-dominant palms: "lontar" (<u>Borassus sundaicus</u>), a sap-producing palm that forms pure stands on highly degraded sites in eastern Indonesia, and "babassu" (<u>Orbignya phalerata</u>), an oil-producing palm that occurs on secondary forest sites in central and northern Brazil. Lontar and the people who use it represent a highly successful example of integration between palm forests and rural communities. In contrast, babassu stands are being eradicated over widespread areas due to conversion of formerly cultivated lands to cattle ranches. The viability of palm forests as a resource seems to depend on whether they occur on sites where competitive and potentially disruptive land uses are practiced. The two examples suggest that use and management of these forests are viable enterprises on climatically and/or edaphically marginal sites.

Based on the scant evidence that currently exists, this paper proposes that palm forests possess a potentially unique pattern of nutrient cycling, which enables them to support relatively productive and stable forms of agriculture as well as to contribute to recovery of disturbed sites.

10.2. INTRODUCTION

Agriculture is a dubious enterprise over widespread areas of the lowland tropics. Sites that are marginal for agriculture occur in humid lowlands subject to high nutrient leaching, weed infestation and pest attack; in dry areas subject to unpredictable rainfall and intensive erosion; and in areas subject to prolonged inundation. If trees are able to grow on such sites, management of native forests is often cited as the most preferable form of land use from an ecological perspective (e.g., 13). From an economic perspective, however, the high biological diversity characteristic of native forests in the lowland tropics often represents a serious obstacle to their long-term management and utilization.

Although high biological diversity is the rule, native forests that are low in diversity are surprisingly common in the tropics. These forests are often dominated by economically important species of palms (Table 1). The "lontar" palm (<u>Borassus sundaicus</u> Beccari), which dominates derived savannas in Indonesia (11), and the "nipa" palm (<u>Nypa</u>

Table 1. Some examples of forest-dominant palm species throughout the world.

Species	Chief Economic Product	Chief Habitat	Distribution
Borassus sundaicus	Sugary sap	Derived savannas	Indonesia
Nypa fructicans	Sugary sap	Saltwater swamps	S.E. Asia-Australia
Mauritia flexuosa	Starch	Freshwater swamps	Northern S. America
Elaeis guineensis	Vegetable oil	Secondary upland forests	W. Africa
Orbignya phalerata	Vegetable oil	Secondary upland forests	Northern S. America

fructicans Wurmb.), which occurs in pure stands in coastal swamps from Southeast Asia to Australia (26, 31), produce edible sap that is marketed and consumed by local inhabitants. The "sago" palm (Metroxylon sagu Rottb.), which abounds in freshwater swamps of Southeast Asia and the Pacific, is an important source of edible starch (35). A counterpart to the sago palm in the Americas is the "buriti" or "moriche" palm (Mauritia flexuosa L.f.), which dominates in freshwater swamps throughout northern South America (16). In addition to plantations established throughout the humid tropics, the oil palm (Elaeis guineensis Jacq.) forms extensive, high-density stands on secondary forest sites in west Africa (39). A counterpart to oil palm in the Americas is the "babassu" palm (Orbignya phalerata Mart.), which flourishes on secondary forest sites in Brazil, Bolivia and the Guianas (2). The preceeding is only a small sampling of economically important palms that are locally abundant. In the Brazilian portions of the Amazon Basin, for example, the list could be expanded to include at least 22 species in 12 genera (Table 2).

Forest-dominant palms often play a crucial role in both market and subsistence economies. These palms provide commercial products such as fibers, waxes, vegetable oils, edible fruits, beverages, palm hearts and flavorings. Products from native palm forests make a substantial contribution to national economies. For example, in 1979 Brazil reported over U.S. $100,000,000 of commerce from the sale of products from six native palm genera (all represented by species that are locally abundant): Astrocaryum, Attalea, Copernicia, Euterpe, Mauritia and Orbignya (18). In addition to market products, palm forests provide an astonishing array of subsistence products, including shelter, clothing, foods, beverages, oils, protein from palm-feeding larvae, charcoal, kitchen utensils, tools, weapons, bait, hammocks, baskets, fishing nets, brooms, ornaments, cosmetics, toys, medicine and magic (26). As shown in the case studies described below, the combination of market and subsistence products obtained from these forests often makes a crucial difference in rural economies.

Table 2. Forest-dominant palm genera of economic importance in Brazilian Amazonia. Numbers in parentheses represent probable number of species.

Genus	Habitat(s) Dominated	Chief Economic Use(s)
Acrocomia (1)	Secondary upland forests	Oil
Astrocaryum (4)	Secondary upland forests, Permanently inundated forests	Oil, edible fruits, fiber
Euterpe (3)	Temporarily inundated forests	Palm heart, flavoring, oil-rich beverage
Jessenia (1)	Temporarily inundated forests	Oil, oil-rich beverage
Leopoldina (1)	Temporarily inundated forests	Fiber
Manicaria (1)	Temporarily inundated forests	Fiber, starch
Mauritia (3)	Temporarily inundated forests	Fiber, starch, edible fruits, flavoring, oil-rich beverage
Maximiliana (1)	Secondary upland forests	Oil
Oenocarpus (3)	Secondary upland forests, Temporarily inundated forests	Oil, oil-rich beverage, flavoring, palm heart
Orbignya (2)	Secondary upland forests	Oil, charcoal
Raphia (1)	Permanently inundated forests	Fiber
Scheelea (1)	Secondary upland forests	Oil

Forest-dominant palm species usually occur on marginal sites that are not suitable for agriculture. For example, most of the genera listed in Table 2 dominate temporarily or permanently inundated sites where agricultural activities are minimal or nonexistent; other genera flourish on sites that have been degraded due to agricultural activities. As described in the case studies below, palms appear to provide an important and possibly unique contribution to site restoration through their role in the recovery of deep soil nutrients. Palm forests also provide important sources of food for animal communities (9, 20, 21, 7). Many animals that are hunted for game appear to thrive in palm forests (e.g., 36, 23, but see 28). Domesticated animals such as chickens and pigs often obtain important nutritional supplements from products harvested from palm forests. These forests may also play an important role in the maintenance of aquatic communities, as evident from the nutritional dependence of the economically important "tambaqui" (Colossoma macropomum) and other fishes on the fruits of the "jauari" palm (Astrocaryum jauari Mart.) in Amazonian Brazil (14).

Palms attain high density populations that are generally self-maintaining and relatively pest free. The long-term management of palm forests is consequently simple, in contrast to more heterogeneous forests in the tropics. The relatively open crowns of palms permit underplanting of crops and formation of multi-levelled agricultural systems (22). In addition to native palm forests, underplanting of crops and pasture is successfully carried out in plantations of coconut (33) and oil palm (15). In multiple-cropping systems involving coconuts, the combination of species with different root architectures appears to enhance utilization of soil resources (17, 29).

Despite the present and potential importance of native palm forests, surprisingly little is know about them. Few published reviews of knowledge concerning these forests exist even on a regional level (but see 11). Information on the species involved and the sites they dominate are only available in widely scattered sources. Maps showing the distribution of palm forests are limited to a few species (e.g., 4) and are confined to a few individual countries. Ethnobotanical studies are fairly detailed (e.g., 6, 38, 34) but rarely contain quantitative economic data (but see 24, 35). Government sources of data on the economic production of native palms are few and often suspect. Although ecological studies of locally abundant palms are increasing (e.g., 5, 20, 37, 28, 1, 32), almost nothing has been published concerning their management (but see 39, 8).

Why are these economically and ecologically important resources so neglected? In my opinion, one of the chief reasons is that the study of native palm forests lies outside traditional research domains. Many foresters have traditionally focused their attention on wood production, whereas agronomists are trained to work with cultivated crops. The substantial contribution of native palm forests to subsistence economies is extremely difficult to quantify, which has probably contributed to the neglect of these resources by economists. As a result, the study of native palm forests has largely been left to economic botanists, whose publications have apparently been ignored by scientists in other fields.

In this paper, I shall not attempt to provide an exhaustive literature review on the use and management of native palm forests. Given the gaps in the literature, such a review would probably be premature at this time. Instead, I intend to provide a summary of two case studies. The first involves the lontar palm (Borassus sundaicus Beccari) in Indonesia and illustrates the potential contribution of palm forests toward the ecological and economic well-being of rural communities. The second is based on the babassu palm (<u>Orbignya phalerata</u> Mart.) in Brazil and illustrates how tight links between palm forests and people are formed and subsequently broken. I believe this comparison will provide an overview of the use and management of these forests, as well as insight concerning future needs for research.

10.3. LONTAR

The source of this case study is James Fox's (11) "Harvest of the Palm", which provides a penetrating analysis of ecological, historical, and economic changes in the outer arc of the Lesser Suunda Islands in Eastern Indonesia. The current ecological setting of these islands is generally characterized by: "...(a) a short, irregular, wet monsoon season with low overall rainfall [usually less than 1,000 mm per year]; (b) a prolonged dry season, often initiated by tropical cyclones and dominated by a dry westerly wind; and (c) impermeable, erosion-prone, margalitic soils that suffer extreme desiccation during the long dry season" (p. 18). When population densities exceed 20 to 25 persons per square kilometer, grazing and shifting cultivation under such extreme conditions have led to widespread environmental degradation and economic impoverishment. Yet three islands (Savu, Roti, and Ndau) with the lowest overall rainfall and highest population densities support communities that are economically prosperous. In contrast to their neighbors, these islands contain extensive stands of the lontar palm, which originate on severely degraded sites and provide the basis for a complex and diversified economy.

The principal element of this economy is a sweet juice that the island inhabitants extract by breeching the inflorescences of the palm. The lontar is dioecious, and both male and female palms produce large quantities of juice during the dry season (April through November). This juice is consumed directly, and during the dry season it constitutes the principal food for the islanders. Because it sours quickly, juice not immediately consumed is cooked to produce a syrup, which can be stored for prolonged periods and subsequently mixed with water to make a sweet drink; this drink provides the normal sustenance when fresh juice is not available. The syrup can also be crystallized to form a brown rock sugar or fermented to make beer. People use this fermented beer as a mash from which they distill gin.

Although palm tapping occurs throughout the dry season, there are two peak periods: one at the beginning of the dry season in April and May, and the other near its end in September and October. The syrup produced during the first period is either sold directly or used for making gin, which is exported to regional population centers. During the second, more intensive tapping period, each household produces most of its annual reserve. In drier years when agricultural production is poor, people utilize the lontar more intensively. The palms have never failed and, in a mere two months (during which agricultural activities are minimal) they provide the island inhabitants with most of their subsistence needs.

The fan-shaped leaves of the lontar provide raw material for a bewildering array of subsistence products, including roof thatch, umbrellas, bucket-like containers, musical instruments, baskets, sacks, saddlebags, mats, fans, toys, hats, belts, sandals, thread for clothing and sails, knife sheaths, rope, bindings, straps, harnesses, bridles, fences, walls, partitions, firewood and coffins. After an islander has died, a lontar leaf is used to form a three-pronged, fork-like object that represents his spirit. Thus it can be said that the people of these islands are fed, equipped, clothed, buried and remembered by the products of their palms.

Lontar supports a variety of other economic activities. Human settlement is invariably concentrated in and about the palm stands, which provide the basis for permanent garden plots. Each year fallen lontar leaves are gathered, laid down in multiple layers over each garden, and burned. The resulting release of nutrients enables these relatively small gardens to obtain greater yields than more extensive areas cleared for shifting cultivation.

Shifting cultivation, which is practiced where lontar stands do not occur, is not only less productive but requires more work. High population densities combined with the extensive nature of shifting cultivation require that agricultural plots be located at considerable distances from human settlements. Protection of these plots against grazing animals is especially difficult. The only available fencing materials consist of stout tree trunks and heavy branches that must be gathered from extensive areas. Construction and maintenance of such fences require one fourth to one third of a farmer's total labor input. In contrast, where lontar abounds the leaves of the palm are used to make fences that are easily constructed and maintained.

Because the permanent gardens associated with lontar require relatively little investment of time and effort, rural inhabitants are able to pursue alternative economic activities. One of the most common of such activities is pig-raising. Virtually all rural households own up to seven or eight pigs, which are fed fresh lontar juice, as well as the residue and spill from syrup cooking. These pigs--which effectively convert calories from lontar into protein--are raised exclusively for domestic consumption. An additional activity is honey-gathering, which is especially successful because the sap-producing lontar stands support large bee populations. Due to the abundant supply of sugar foods, the islanders usually sell honey as a valuable export commodity.

Relative freedom from agriculture permits people who live near lontar stands to engage more intensively in other subsistence activities such as fishing and seaweed-gathering. The former provides an additional protein source and the latter an important dietary supplement in regions where there is a distinct lack of vegetable foods.

In summary, stands of the lontar palm form spontaneously on severely degraded sites. The stands substantially increase the carrying capacity of such sites by providing a diverse array of products and supporting a variety of additional economic activities. The highly diversified network of subsistence and market economies associated with lontar provides numerous options for rural inhabitants should one component fail. These economies are eminently successful, as evident from their current spread to other islands such as Timor, where natural stands of lontar are beginning to form. Intentional planting of the palm has already begun on the island of Savu, and ample potential would appear to exist for more widespread use of the palm for reforestation of degraded landscapes.

10.4. BABASSU

Stands of the babassu palm occur over widespread areas of central and northern Brazil, attaining their greatest coverage (102,970 km^2) in the state of Maranhao (2)[1]. In contrast to lontar, babassu flourishes in relatively fertile and moist sites that have moderate to high potential for agriculture (10). Rainfall in the babassu zone of Maranhao ranges from 1,200 to 2,000 mm per year, over 80% of which falls during a 5-6 month wet season. The predominant soils are characterized by relatively high (50%) base saturation and generally excellent structure.

Local inhabitants living near stands of babassu and the closely related "cohune" palm (Orbignya cohune (Mart.) Dahlgren) in Central America consider these palm to be reliable indicators of agriculturally "good" soils (12, 27). Two alternative explanations could account for this observation: either these species of Orbignya are intolerant of agriculturally "poor" soils or else they are effective in enhancing soil quality. Furley (12) suggested the latter and proposed an interesting mechanism. Species of Orbignya--well as other palm genera represented by abundant species (e.g., Attalea, Maximiliana, Scheelea, Sabal)--possess a curious mode of germination in which the apical meristem is pushed below the soil surface. This mechanism, known as "cryptogeal germination" (19), causes the stem to arise from below ground. After a stem of the cohune palm dies, it quickly decays, leaving a large hole (ca. 0.5 m in diameter and 0.5-1.0m in depth) that eventually collapses. This process of soil turnover probably contributes substantially to increased recycling of deep soil nutrients and improved soil structure under stands of Orbignya.

Regardless of their origin, the fertile and well-structured soils associated with babassu enabled agriculture to flourish in Maranhao. Accelerated influx of settlers since the 1950s has resulted in widespread removal of primary forests by shifting cultivators. Babassu, a common component in primary forests, completely dominates cleared sites due to its cryptogeal mode of germination, which permits juvenile palms to survive the repeated cycles of cutting and burning associated with shifting cultivation. During the subsequent fallow, these juveniles are simultaneously released and form dense, uniform stands.

Forests of babassu originate and are maintained by human activities. People, in turn, depend on these forests for food, fuel, shelter and income (25). Subsistence products from babassu leaves include baskets, mats, fans, sieves, twine, bird cages, hunting blinds, animal traps, thatch, laths and rails. The various components of babassu fruits provide rural inhabitants with cooking oil, snack food, flour,charcoal, fuel for lamps, animal feed, game attractants, fish and shrimp bait, various kinds of medicine and handicrafts. In addition, manual extraction of kernels from the extremely hard fruits of babassu constitutes a cottage industry involving over 500,000 families in Brazil. During the 3-month peak of the babassu harvest (October-December), May et al. (25) found that selling of kernels contributed an average of 40% of total monetary income for households in three rural districts ("municipios") in Maranhao.

Not only is babassu an integral part of subsistence and market economies, but of major land uses as well. The palm's current dominance of the landscape is a testimony to its remarkable adaptability to shifting cultivation. Under this system, a site is periodically cleared, burned, cropped and then left in fallow. If the fallow period is long, sufficient fuel accumulates in the form of biomass to permit the hot fire necessary for recycling nutrients and weed control. Shifting cultivators operating in areas dominated by babassu traditionally carry out only moderate

---

[1]Reference 2 is the principal source utilized this summary on babassu; other sources are cited where relevant.

thinning of the stands; most of the fuel utilized for the burn is in the form of palm leaves. In a dense babassu stand, Anderson (1) found that the dry weight of living leaves was 69.1 tons per hectare (Table 3); the figure for leaves in other tropical dry forests rarely exceeds 10 tons per hectare. The exceptionally high leaf biomass in babassu stands permits shifting cultivators to obtain sufficient fuel by cutting leaves instead of stems. Cutting leaves also serves to reduce shading of cultivated crops during the subsequent growing season. In addition to high leaf biomass, the same babassu stand produced 16.8 tons of leaves per hectare per year (Table 3), which is an order of magnitude greater than produced by most tropical dry forests. As defoliation at sufficiently wide intervals does not appear to reduce leaf productivity, a burned stand consequently requires about four years to recover its full biomass of leaves. Given these fallow periods, a moderate to high density of mature palms can be maintained under shifting cultivation, providing both fuel for the cropping cycle and valuable resources during the fallow.

Despite its apparent stability, shifting cultivation in portions of the babassu zone in Maranhao is currently breaking down. This breakdown is primarily caused by widespread expansion of cattle ranches into formerly cultivated areas (3). With less land available for farming, shifting cultivators are compelled to reduce their fallow periods. In the Mearim valley of central Maranhao, for example, the average fallow period had fallen to 3.3 years by 1980 (3). To obtain sufficient fuel for burning, shifting cultivators are consequently compelled to thin the babassu stands more intensively, thus destroying the very resource upon which they depend for income and subsistence products.

With widespread degradation of shifting cultivation sites in areas such as the Mearim valley, rural poor are becoming increasingly dependent on the stands of babassu growing in planted pastures. Ranchers, in turn, frequently perceive the gatherers of babassu fruits as interfering with pasture management. Fruit gatherers are accused of starting wildfires,

Table 3. Biomass and productivity of babassu palms in a secondary forest site at Lago Verde, Maranhao, Brazil (1).

Component	Biomass $(t \cdot ha^{-1})$	Productivity $(t \cdot ha^{-1} \cdot yr^{-1})$
STEMS (TOTAL)	81.4	4.8
LEAVES		
Leaflets	19.0	4.7
Leaf axes	50.1	12.1
Total	69.1	16.8
REPRODUCTIVE STRUCTURES	–	1.5
Ancillary Structures	–	1.7
Fruits	–	3.2
GRAND TOTAL	130.5	24.7

cutting fences, and leaving behind broken fruit husks that damage the hooves of cattle. To rid themselves of these incursions, ranchers frequently prohibit or restrict access to their property and are increasingly engaged in clearcutting of the stands.

Clearcutting has probably eliminated less than 5% of the total area of babassu stands of Maranhao (Anderson and May 1985). In regions such as the Mearim valley, however, the rate of deforestation appears to be increasing rapidly. For people dependent on the palm as a principal source of income and subsistence products, the potential consequences of an increasing rate of deforestation are grim indeed.

10.5. DISCUSSION

The two case studies described above illustrate how palm forests can play an integral role in rural economies, and how that role can be undermined. Whereas the economies associated with lontar are successfully spreading, those associated with babassu are breaking down in certain areas. The key to lontar's comparative success appears to be its occurrence on extremely degraded sites that are agriculturally marginal. In contrast, babassu stands occur on relatively humid, fertile sites with moderate to high potential for agriculture. On such sites a competitive land use (conversion to planted pastures) is leading either directly or indirectly to eradication of palm forests. In drier areas of the babassu zone of Maranhao, where widespread pasture conversion has not occurred, palm forests and their associated economies continue to flourish (3).

The two case studies therefore suggest that use and management of native palm forests are a more viable enterprise on agriculturally marginal sites (Table 4). This is generally true for agroforestry systems throughout the world, which seem to be best suited to sites where more capital intensive land uses are not competitive.

Both lontar and babassu stands form spontaneously and require little or no management. Both provide a varied mixture of subsistence and market products, which appears to be crucial for maintaining economic well-being in rural areas. In addition to products derived directly from these palms, stands of lontar and babassu support other economic activities. In particular, both appear to have crucial roles in maintaining stable, productive forms of agriculture. In the case of Orbignya and several other palm genera, an unusual mode of germination may ultimately contribute to recycling of deep soil nutrients and improved soil structure (12). In addition, the leaves of both lontar and babassu are actively utilized to promote soil fertility. I believe that this latter observation points to a general mechanism operating in palm forests.

The figures obtained by Anderson (1) for leaf biomass (69.1 tons per hectare) and leaf productivity (16.8 tons per hectare per year) are among the highest figures recorded for a forested site, whether native or planted. The only forested site that I have found with comparable leaf biomass and productivity is a plantation of oil palm in Malaysia (30). In both babassu and oil palm the lion's share of total biomass and productivity are allocated to leaves. If one can generalize from two examples, I suspect that this is a general feature in palms and distinguishes them from dicotyledonous trees, in which relatively higher amounts of biomass and productivity are allocated to woody support structures such as stems and branches.

This fundamental difference could have profound implications for systems of shifting cultivation. In forests dominated by babassu, for

Table 4. Comparison of two forest-dominant palm species.

	Lontar (_Borassus sundaicus_)	Babassu (_Orbingnya phalerata_)
Principal Reference	Fox (11)	Anderson & Anderson (2)
Distribution	Indonesia	Bolivia, Brazil, Guianas
Annual Rainfall	500-1,000 mm	1,000-2,000 mm
Soils	Impermeable, desiccated, agriculturally "poor"	Permeable, moist, agriculturally "good"
Principal Subsistence Uses	Foods, shelter, fuel, numerous implements, construction materials, clothing, magic	Foods, shelter, fuels, numerous implements, construction materials handicrafts, medicines
Other Economic Activities Supported	Permanent cultivation, gathering, animal domestication	Shifting cultivation, hunting, animal domestication
Future Prospects	Excellent	Dubious

example, palm leaves provide most of the fuel for the burn, thinning of stands is minimal, forest cover is maintained, and a stable forest resource is available during the fallow. In forests dominated by dicotyledonous trees, stems and branches provide most of the fuel, which means that these forests much be clearcut or drastically thinned before the site can be burned. As a result, forest cover and forest resources are temporarily lost; both of these losses are likely to lead to numerous additional stresses on shifting cultivation systems.

Although the evidence is admittedly thin, these considerations lead me to hypothesize that, other factors being equal, shifting cultivation is a more successful long-term enterprise in palm forests than in forests dominated by dicotyledonous trees. In the former, shifting cultivators can obtain sufficient fuel for burning without destroying the overall forest structure. Furthermore, high allocation of biomass and productivity to leaves--which I suspect is a general characteristic of palm forests--provides a more rapid mechanisms for recycling of nutrients in these forests. Gathering these leaves and concentrating them on selected sites serves as the basis for more permanent forms of agriculture, such as is practiced in backyard gardens under lontar stands in Indonesia.

10.6. CONCLUSIONS

The foregoing discussion is highly speculative and reveals some of the gaps in our knowledge concerning the use and management of native palm forests. Although the literature is extremely deficient, the studies that do exist indicate that native palm forests have a number of attractive characteristics. They form spontaneously, require little or no management, promote site recovery, provide an exceptional variety of market and subsistence products, and support alternative economic activities. Their potentially unique role in nutrient recycling is a virtually unexplored question that is ripe for investigation. Integrated studies of native palm forests from a biological, social, and economic perspective are essential to understand how these forests are currently used and managed, as well as to assess their potential in alternative land uses such as reforestation.

ACKNOWLEDGMENTS

Field research on babassu was supported by a grant from the U.S. Man and the Biosphere (MAB) Program. Support while preparing this paper was provided by the Ford Foundation and the Conselho Nacional de Desenvolvimento Cientifico e Technologico (CNPq).

REFERENCES

1. Anderson, A. B. 1983. The biology of Orbignya martiana (Palmae), a tropical dry forest dominant in Brazil. PhD Dissert, Univ Florida, Gainesville, FL.

2. Anderson, A. B. and E. S. Anderson. 1983. People and the palm forest: Biology and utilization of babassu forests in Maranhao, Brazil. Final report, US Dep Agric, For Serv, Consort Study Man's Relationship with Global Environ, Washington, D.C.

3. Anderson, A. B. and P. H. May. 1985. A palmeira de muitas vidas. Ciencia Hoje 4:58-64.

4. Anonymous. 1981. Mapeamento das ocorrencias e prospeccao do potencial atual do babacu no Estado do Maranhao. Companhia de Pesquisa e Aproveitamento de Recursos Naturais (COPENAT) and Fundacao Instituto Estadual do Babacu (INEB), Sao Luis, Brasil.

5. Bannister, B. A. 1970. Ecological life cycle of Euterpe globosa Gaertn. In: H. T. Odum and R. F. Pidgeon (eds), A tropical rain forest. U.S. Atomic Energy Commission, Oak Ridge, TN.

6. Barrau, J. 1959. The sago palm and other food plants of marsh dwellers in the South Pacific islands. Econ Bot 13:151-162.

7. Bradford, D. F. and C. C. Smith. 1977. Seed predation and seed number in Scheelea palm fruits. Ecology 58:667-673.

8. Calzavara, B. B. G. 1972. As possibilidades do acaizeiro no estuario amazonico. Boletim da Fundacao de Ciencias Agrarias do Para (FCAP), no. 5, Brasil.

9. Costa Lima, A. M. 1967-68. Quarto catalogo dos insetos que vivem nas plantas do Brasil, seus parasitos e predadores. Ministerio de Agricultura, Rio de Janeiro, Brasil.

10. EMBRAPA. 1981. Mapa de solos do Brasil - Escala 1:5.000.000. Servico Nacional de Levantamento e Conservacao de Solos, Rio de Janeiro, Brasil.

11. Fox, J. J. 1977. Harvest of the palm. Harvard Univ Press, Cambridge, MA.

12. Furley, P. A. 1975. The significance of the cohune palm, Orbignya cohune (Mart.) Dahlgren, on the nature and in the development of the soil profile. Biotropica 7:32-36.

13. Goodland, R. 1980. Environmental ranking of Amazonian development projects in Brazil. Environ Conserv 7:9-26.
14. Goulding, M. 1980. The fishes and the forest. Univ California Press, Berkeley, CA.
15. Hartley, C. W. S. 1977. The oil palm. Longman, London.
16. Heinen, H. D. and K. Ruddle. 1974. Ecology, ritual, and economic organization in the distribution of palm starch among the Warao of the Orinoco Delta. J Anthropol Res 30:116-138.
17. Huck, M. G. 1983. Root distribution, growth, and activity with reference to agroforestry. In: P. A. Huxley (ed), Plant research and agroforestry, pp 527-542. International Council for Research in Agroforestry (ICRAF), Nairobi, Kenya.
18. IBGE. 1981. Producao extrativa vegetal--1979. Fundacao Instituto Brasileiro de Geografia e Estatistica (IBGE), Vol 7, Rio de Janeiro, Brasil.
19. Jackson, G. 1974. Cryptogeal germination and other seedling adaptations to the burning of vegetation in savanna regions: The origin of the pyrophytic habit. New Phytol 73:771-780.
20. Janzen, D. H. 1971. The fate of Scheelea rostrata fruits beneath the parent tree: predispersal attack by bruchids. Principes 15:89-101.
21. Janzen, D. H. 1972. Association of a rain forest palm and seed-eating beetles in Puerto Rico. Ecology 53:258-261.
22. Johnson, D. 1983. Multi-purpose palms in agroforestry: A classification and assessment. Intern Tree Crops 2:217-244.
23. Kiltie, R. A. 1981. Distribution of palm fruits on a rain forest floor: Why white-lipped peccaries forage near objects. Biotropica 13:141-145.
24. Martin, A. 1956. The oil palm economy of the Ibidio farmer. Ibadan Univ Press, Nigeria.
25. May, P. H. Anderson, A. B., Balick, M. J. and J. M. F. Frazao. 1984. Subsistence benefits from the babassu palm (Orbignya martiana). Econ Bot 39:113-129.
26. Moore, H. E. 1973. Palms in the tropical forest ecosystems of Africa and South America. In: B. J. Meggers, E. S. Ayensu and W. D. Duckworth (eds), Tropical forest ecosystems in Africa and South America: a comparative review, pp 63-88. Smithsonian Inst Press, Washington, D.C.
27. Moran, E. F. 1981. Developing the Amazon. Indiana Univ Press, Bloomington, IA.
28. Myers, R. L. 1981. The ecology of low diversity palm swamps near Tortuguero, Costa Rica. PhD Dissert, Univ Florida, Gainesville, FL.
29. Nair, P. K. R. 1983. Agroforestry with coconuts and other tropical plantation crops. In: P. A. Huxley (ed), Plant research and agroforestry, pp 79-102. International Council for Research in Agroforestry (ICRAF), Nairobi, Kenya.
30. Ng Siew Kee, S. and P. DeSouza. 1968. Nutrient contents of oil palms in Malaysia. II. Nutrients in vegetative tissues. Mal Agr J 46:332-402.
31. Paivoke, A. E. A. 1984. Tapping patterns in the nipa palm (Nypa fruticans Wurmb.). Principes 28:132-137.
32. Pinero, D., Martinez-Ramos, M., Mendoza, A., Alvarez-Buyilla, E. and J. Sarukhan. 1986. Demographic studies in Astrocaryum mexicanum and their use in understanding community dynamics. Principes 30:108-116.

33. Plucknett, D. L. 1979. Managing pastures and cattle under coconuts. Westview Press, Boulder, CO.
34. Putz, F. E. 1979. Biology and human use of _Leopoldinia piassaba_. Principes 23:149-156.
35. Ruddle, K., Johnson, D., Townsend, P. K. and J. D. Rees. 1978. Palm sago: a tropical starch from marginal lands. Univ Press Hawaii, Honolulu, HI.
36. Smith, N. 1974. Agouti and babassu. Oryx 12:581-582.
37. Vandermeer, J. H., Stout, J. and G. Miller. 1974. Growth rates of _Welfia georgii_, _Socratea durissima_ and _Iriartea gigantea_ under various conditions in a natural rainforest in Costa Rica. Principles 18:148-154.
38. Wilbert, J. 1976. _Manicaria saccifera_ and its significance among the Warao Indians of Venezuela. Bot Museum Leaflets Harvard Univ 24:275-335.
39. Zeven, A. C. 1967. The semi-wild oil palm and its industry in Africa. PhD Dissert, Univ Wageningen, Wageningen, Netherlands.

11. LIVING FENCES IN TROPICAL AMERICA, A WIDESPREAD AGROFORESTRY PRACTICE

Gerardo Budowski

Head, Program for Natural Resources and Quality of Life, University for Peace, Costa Rica

11.1. ABSTRACT

The practice of using living fence posts to attach rows of barbed wire is widespread in tropical America although related scientific knowledge is relatively scarce. Besides holding wire, live fences produce fuelwood, fodder, and food, and act as windbreaks and protection for wildlife, but the greatest benefit is derived from the use of branches to establish more fences or to "fill in" old fences. Many trees are used, depending on ecological zones, availability of large cuttings for planting, and special needs dictated by preferences and beliefs of the farmers. Planting practices, studied in detail in Costa Rica, also vary. Advantages and drawbacks of living compared to non-living wood fences are discussed. Some speculations of future prospects and the possible involvement of scientists are advanced.

11.2. INTRODUCTION

Living fence posts are a characteristic feature of the landscape of many tropical American countries from sea level to well above 2500 meters, and from relatively dry environments to some of the very wettest areas (over 4000 mm annual rainfall). As here described and analyzed, living fences refer only to those that are established by planting large cuttings (usually about 2.5 m long and from 8-20 cm in diameter), that easily produce roots and on which several strings (usually 3) of barbed wire are attached with the obvious purpose of keeping livestock in or out. Besides fencing, many other possible benefits are derived from the trees, including the production of various goods and services. Although publications in this area are few, a pioneer research effort has been carried out in recent years at the Tropical Agricultural Center for Research and Training (CATIE) in Turrialba, Costa Rica, where a number of M.S. theses have been produced. The rooting ability of some newly planted fences was examined as early as 1960 (14). More recently, studies on the best frequency of pruning _Gliricidia_ _sepium_ for fodder (3) and current management practices of farmers for preparing _Gliricidia_ cuttings and planting (1) showed other promising lines of research. Moreover, various CATIE staff members have been producing information on species selection, productivity, socio-economic implications, advantages and drawbacks of live fences compared with "dead" fences (2, 6, 8, 17). In an important paper, Sauer (18) made an eloquent plea for the devotion of more attention to living fences and for preventing their disappearance because of pressures for mechanization or substitution by non-living fences.

This paper attempts to summarize the present information. Part of the material has been taken from a forth-coming book on living fences (7).

11.3. SOME HISTORY AND SOME ODDITIES

Accounts of living fences can be found in early literature, particularly contributions by botanists who notice that certain species were propagated by large cuttings to establish fences or hedges. The word "fence" is here used to indicate a single row usually used as support for wire, while hedges may be much thicker and most of the time include a number of species, including trees, shrubs and smaller plants, usually without wires.

Early accounts of living fences and hedges include those of Burgos (9) in Peru, Crane (12) in Cuba, Bond (5) in Nigeria for hedge plants, Howes (13) for the tropics in general, and Lozano (14) in Costa Rica. Most of these studies described species and some management practices but without going into much detail.

Living fences (as well as hedges and windbreaks) were probably first incorporated into an agroforestry classification scheme in 1979 by Combe and Budowski (11). They have since been incorporated into the wide definition of agroforestry proposed by ICRAF.

It is significant that documentation of the planting of living fences is not found among any agricultural or forestry statistics, nor among those for animal husbandry although it is here where the widest use is made. I have often asked the question of Latin American audiences: "Which are the most commonly planted trees in rural landscapes?" Rarely does the audience think of the planting of live fences. The answer, I would propose, is <u>Gliricidia</u> <u>sepium</u>, <u>Bursera</u> <u>simaruba</u>, <u>Spondias</u> <u>purpurea</u> or <u>Erythrina</u> <u>berteroana</u>, the four most common species I estimate to have been planted in Central America, northern South America and several Caribbean countries. In some countries where live fences exist, they go unnoticed by foresters and agronomists alike; the practice is virtually ignored by most cattle and grazing specialists. (Metallic or electric fences are preferred, but they are obviously very expensive and few small farmers can afford them.) These attitudes are symptomatic, since living fences have been very important to local farmers probably forever. An insight into this paradox would be a useful contribution to any effort in improving rural development programs.

11.4. WHY LIVE FENCES ARE SO POPULAR

Live fences when compared to "normal" wooden fences have both advantages and drawbacks. An attempt has been made (Table 1) to tabulate those in a comparable way, but without pretense for quantification (a task that remains to be done). Some factors cannot easily be quantified. For instance, for farmers who can afford it, there appears to be some prestige in displaying metallic or electric fences, and, if wooden fences are used (preserved or not), they are often painted white.

Revisions to Table 1 could be to subdivide according to species and ecological or socio-economic conditions. For instance, what may be advantages of a live fence in a wet area (e.g., a large use of water and a superficial root system) may be disadvantages in a dry area when the roots compete for water. Establishment will vary widely according to availability and accessibility of raw materials for fences, costs of labor, demand for establishing new fence posts by another farmer, and so on.

Some years ago in Costa Rica, it was customary for most farmers who owned live fences to allow other farmers to cut branches as raw material for establishing their own fences; now, a price is usually exacted. In a survey done in 1985 in Honduras, a few kilometers south of San Pedro Sula, a stake of Gliricidia was worth US $0.50, cut, next to the fence where it came from, close to the road and ready to be carried off. In Costa Rica in 1985, stakes of Erythrina berteroana in the Turrialba valley were sold at between US $0.16 and $0.32 depending on size and condition. One old "fence arranger" would charge nothing to prune other farmer's fences--an operation that needs to be done usually once a year--but benefited when another farmer contracted him to install a new live fence on his farm. From the previous "free" pruning he obtained the stakes for his new contract, and this became his income as a professional "fence pruner". Since he had earned a well established reputation for pruning, preparing stakes and planting them, he was very much in demand in the region.

If cut stakes are not sold for new fences there are other way of taking advantage of them, as well as the trimmings resulting from the pruning. For instance, the stakes can be used to "fill in" the fence from which they originated, a very common practice and a logical one: because the wire is better held in place and becomes less prone to be dismantled, the lifetime and usefulness of the fence are increased. Many other products may be derived from live fences: for instance, forage for cattle or small animals (goats, rabbits, chickens), as with Gliricidia sepium or Erythrina spp.; ornamental cuttings to be sold or even exported, as with Yucca elephantipes or Dracaena spp.; colorful seeds to be sold in curio shops (e.g., Erythrina sp.); or fuelwood, particularly the thicker branches of Gliricidia.

Edible fruits and flowers can become important. "Jocote" (Spondias purpurea) is locally very important in all Central American countries. There are even selected varieties (e.g., "Jocote tronador" in Guatemala). When in fruit (October-December), the fruits can be found in markets and fruit stands throughout the country along roads close to the live fences that produced them. Large flowers of "itabo" or "izote", Yucca elephantipes, can be sold from $0.50 to $1.00 each in markets. In Costa Rica, they are prominently displayed from January to March and several recipes are found in cook books (mostly as a vegetable or mixed with eggs).

Of the 92 species identified in living fences in Costa Rica (7), 35 are propagated almost exclusively by large stakes, 20 are proven nitrogen fixing trees, 41 are used as firewood (although of variable quality depending on the species and the age of the branches), 24 produce edible fruits or flowers, and 21 produce forage for cattle (mostly polygastrics), rabbits or chicken. For all 92, there is at least one additional product besides fencing, such as honey, medicine, wind protection, nesting sites for important bird and animal species, support for orchid growing, enrichment of soil (particularly by N fixing species), and cultivation for sale as an ornamental.

There are also some drawbacks to living fences. Pruning is indispensable or the fences become trees and their trunks become too thick to be cut, and their branches may produce too much shade. Although infrequent, it is possible to see these fences "having gone wild" in certain abandoned farms. Some fences, particularly old ones or when pruning has not been carried out for a few years, tend to grow very fast in diameter and "swallow" the wire, which may then rust and break.

Table 1. Comparisons between live and wooden (dead) fences (the latter either naturally durable or treated with preservatives). Both methods use attached barbed wire and are used principally to prevent trespass by cattle (6).

Factor	Live fence	Wooden (dead) fence
Choice of species	Depends on ecological conditions	Many possibilities: depends on availability
Cost	Relatively low, or free	Relatively high
Banding of post before placing	Needs careful preparation, transport and storage	No special care required
Placing in soil	Needs care, adequate soils	Soil not limiting
Placing of barbed wire	Special techniques in some aspects	Some skills required
Initial maintenance	Necessary, requires protection against some animals	None, in some cases needs fire protection
Survival	Losses possible	Usually no losses if well-treated
When to place wire	Usually when well-anchored	Immediately
Increase of post density along fence	Easy and cheap	Easy but expensive
Durability	Usually very long	Variable, limited according to treatment and species
Organic matter production	Varies with species	None
Nitrogen fixation	Possible in some species	None
Effect on soil fertility	Generally beneficial	None
Erosion control	Can be effectively used as barrier	None

Competition for water and nutrients and light with nearby crop	Does exist but varies according to system; organic matter production compensates	None
Protection of crops and/or animals against wind	Effective but varies according to species, height, density	Little or none
'Horizontal' rain (fog drip) from moisture laden winds	Possible	None
Toxic effects	Possible (allelopathy)	None (except when some preservatives are used)
Toxic fauna	Can be sheltered	None (except termites in some cases)
Beneficial fauna	Provides shelter and food (for example for birds, honey bees)	Few, if any
Additional economic products	Many, such as food, feed, medicinal products, also firewood, posts and more live fence posts	None
In case removal is necessary	Difficult and costly	Relatively easy
Labor for management	Periodical pruning is necessary; skills required	Skills also required to place posts and wires and replace them
Acceptance by farmers	Very popular among poorer farmers	Depends on income; more affluent farmers tend to avoid live fence posts
Special limitations	Disliked by aerial crop pesticide sprayers	Firebreaks must be kept clean during fire season
Aesthetics appreciation	Depends on management and cultural background	Depends on investment possibilities and cultural background

11.5. ADAPTABILITY AND AVAILABILITY OF FENCE POSTS IN VARIOUS COUNTRIES AND CLIMATIC ZONES

Living fence posts are found from sea level to about 3000 m. The 92 species so far identified--and this number is growing--range through the following life zones (_sensu_ Holdridge) in Costa Rica: tropical dry, tropical moist, tropical wet, premontane moist, premontane wet, premontane rain, lower montane moist, lower montane wet, and lower montane rain; at least one species, _Drimys winterii_, also goes into the montane belt.

There are few comparative studies from other countries. In trips to Ecuador, Peru, Guatemala, and Mexico, I have found many species not recorded in Costa Rica. Some of these species are nitrogen fixing, produce good fodder, and can easily be propagated by large stakes on which wire can be attached.

The choice of species usually depends on availability of stock; one row of fence usually produces more fences and, if popular, a whole area may soon have the great majority of its fences made up by that species. Thus, in El Salvador, _Jathropha curcas_ has been identified as the most popular species in certain areas, perhaps only because it is widely available. In tropical dry and very dry life zones (for instance, in central west Nicaragua), _Bursera simaruba_ is most popular.

The most widespread species of all is very likely _Gliricidia sepium_, which has many names in Spanish: "madre de cacao" (mother of cacao), "madregao" (a corruption), or "madero negro" in Costa Rica; "mataraton" (mice killer, because of toxic properties of leaves) in Venezuela and Colombia; "bala" in Panama; "pinon florido" in Cuba (where it is not native); "pinon cubano" in Dominican Republic (since it was probably brought to that country from Cuba); "Gliceridia" in the French Antilles (an amusing corruption of the genus); "lila etranger" in Haiti; and so on. It grows from sea level to about 2500 m, from dry regions with a 5-8 month dry season (where it is native), to very wet regions (rainfall sometimes above 5000 mm), where it grows very fast and produces a large amount of biomass. At present it should be rated as the most common live fence to which barbed wire is attached.

Another very versatile species is _Erythrina berteroana_. Like _Gliricidia_, it is a nitrogen fixing species and produces a high quality fodder. It grows from sea level to about 2500 m, but only in relatively moist environments. Some other species, such as _Erythrina costaricensis_, have similar requirements to _E. berteroana_ but are presently much more restricted in their distribution. _Bursera simaruba_ and _Spondias purpurea_ are well adapted to areas where there is a long dry season (lasting up to 8 months). _Bursera_ is rarely seen as a fence in wet areas, although it does grow well and fast there. It has the reputation of having the wire penetrate the bark when it grows too fast ("swallowing the wire"). _Spondias purpurea_ is occasionally planted in wetter areas, but, unless there is a marked dry season, it will produce few, if any, fruits.

11.6. PRUNING

Pruning is indispensable in order to keep a live fence in good shape. The operation is usually carried out once a year and involves a series of practices that vary according to the purpose, the uses of the pruned material, and the personal preferences and experiences of the pruner. Native farmers often prune according to moon phase (waning moon is usually preferred), which also applies to the planting of stakes. Pruning is always carried out by "machete", from the ground or from a small ladder. The following is offered as a summary of pruning objectives.

11.6.1 Selection of stakes for new plantings

For new fences, stakes between 8-15 cm diameter at the base are preferred, usually taking around one to two years to develop. This sounds surprisingly fast, but such branches are often borne by vigorously coppicing trees that may have acquired a diameter of 20-30 cm and a well developed root system, and have encountered little competition from neighboring trees. Hence, when a fence is pruned on a yearly basis, some of the best coppicing branches are often left to grow for another year to be "harvested" for new fences. Obviously, growth of a specific shoot depends on the species, rainfall, soil conditions and other factors.

11.6.2. Filling in old fences.

This is an extremely common practice. For this purpose, stakes of a smaller diameter than for new fences are often used. The wires do not need to be attached immediately but rather a few months later when stakes have taken root well. Two or even more species are often found along the same fence, and it generally turns out that one species was planted for the original row at wide spacing while the second species was planted later to fill in. There are many variations, and I have heard many heated discussions among planters as to how good various mixtures are.

11.6.3. Harvesting fuelwood

The best fuelwood from, for instance, Gliricidia sepium or Diphysa robinioides (another N-fixing legume called "guachipelin" in most of Central America and "macano" in Panama), comes from older branches where a good amount of hardwood has been produced. Hence, pruning for stakes is often combined with harvesting for fuelwood, when side branches, tops and crooked branches are cut and arranged in bundles. There have been some studies on fuelwood productivity of fence species (e.g., 15, 17, 10). In one study (17) fence rows were measured that provided 13.4 and 17.6 metric tons of fuelwood (green weight) per kilometer of fence. Both were harvested when the branches were 8 and 13 months old and, in addition to fuelwood, produced 1580 and 1000 stakes, respectively, for establishing new living fences. In Costa Rica, Beliard and Mora (4) produced a double entry volume table to predict both wood and forage volumes and green and dry weights of Gliricidia based on diameter at the base and length of the branches. This table has been successfully used in other areas with similar conditions and has provided an easy instrument to measure productivity, if appropriate adaptations for different sites are made.

11.6.4. Harvesting for fodder

Although farmers have been feeding Gliricidia to cattle, chickens, rabbits, and even swine, little is known about the long-term effects of feeding it to animals. Usually it is a relatively small proportion of all food intake, even though the leaves contain about 25% protein. Baggio (1) reports that in one farm in Liberia, Costa Rica, Gliricidia amounted to 15% of the total input for 300 swine. It more commonly provides about 25%

of the input to cattle, as in farms of the Guanacaste area of Costa Rica. Normally, the dry weight of leaves and young branches is about 23-26% of green weight, crude protein varies from 20-27% for leaves to somewheat less for small branches, and crude fiber varies from 5.8% (very young leaflets) to 21.5% for adult leaflets (16).

11.6.5. Controlling phenology

Live fences, like normal trees, produce flowers and fruits, and many species are naturally deciduous or lose their leaves because of a dry season. _Gliricidia_, which is naturally distributed in relatively dry lowland areas of Central America, still loses its leaves in wet environments. Pruning may change phenology by precluding flowering and fruiting as well as by delaying loss of leaves. This is a very interesting observation that deserves greater attention. For example, Beliard (3) used several frequencies of pruning to determine that a 6 month interval for forage was quite appropriate but that 3 months was too short to allow biomass recuperation. Naturally, if a 2 year cycle is induced for some branches, the completion of the phenological cycle seems unavoidable. But combinations are possible: e.g., the best 2 or 3 branches may be left for stakes, and the rest periodically pruned.

11.7. COST OF ESTABLISHMENT OF A LIVE FENCE WITH BARBED WIRE

OTS and CATIE (16) conducted a cost analysis of a living fence at Hojancha in the Nicoya Peninsula of Costa Rica (Table 2). In this system, non-living corner posts are first set in. These are made of a very

Table 2. Cost of establishment of 1 km of a live fence at 1.6 m spacing.

Materials	Quantity	Cost per unit (US$)	Total cost (US$)
corner posts	60	1.00	60.00
living posts (stakes)	625	0.33	206.00
barbed wire (3 lines)	4000 m	20/400 m	200.00
Staples (to attach barbed wire)	5 kg	0.76/kg	3.80
Labor	20 days	4.2/day	84.40
		total	$ 554.20

*At the time of the survey $1.00 U.S. = 50 colones approx.

durable wood and are used to "train" the wires. The price of $1.00 US is considered rather inexpensive. For 1.5 km of fencing, 625 stakes are used, about 2.6 m long and with a basal diameter of 6 cm. The $0.33 unit cost includes their cutting, shaping and transport. This amount ($206.00) is probably low because it assumes stakes were harvested from other nearby established live fences. Planting costs are included under "labor". The analysis showed that the living posts amounted to about 35% of the total cost, and was about the same, per km, as the cost of the barbed wire. I am not aware of other such analyses, but conclude that, with all their other benefits, living fences are probably very cost effective.

11.8. FUTURE PROSPECTS

This brief recounting of experiences concerning a very old, yet little known, practice indicates broad opportunities for scientific involvement. Above all, much empirical knowledge has been accumulated by peasants throughout the Latin American tropics. To retrieve such knowledge, careful reviews of farmers' practices need to be made. This by itself would yield valuable results.

With a decreasing supply of naturally resistant wooden posts, formerly cheap and easily available, and the prohibitively high cost of artificially treated posts, living fences are bound to become more common. If one adds the other secondary products that can be derived, this appears to be a remarkable instrument for improving the quality of life of low-income farmers.

To this must be added the fact that very little effort has been made to genetically improve the most desirable fence post species for vigor, biomass production, pruning response, nitrogen fixation, form, and so on. An important factor that lends itself to genetic selection is the relationship between wood and leaves, which could be modified depending on whether stakes, fuelwood or a large amount of leaves are desired.

REFERENCES

1. Baggio, A. J. 1982. Establecimiento, manejo y utilizacion del sistema agroforestal cercos vivos de Gliricidia sepium (Jacq.) Steud., en Costa Rica. MSc Thesis , CATIE/Univ Costa Rica, Turrialba, Costa Rica.
2. Beer, J. 1985. Experiences with fence line fodder trees in Costa Rica. In: Proc Seminar Adv Agrofor. CATIE, Turrialba, Costa Rica.
3. Beliard, C. A. 1984. Produccion de biomasa de Gliricidia sepium (Jacq.) Steud. en cercas vivas bajo tres frecuencias de podas (3.6 y 9 meses). MSc Thesis. CATIE/Univ Costa Rica, Turrialba, Costa Rica. 97 p.
4. Beliard, C. A. and E. Mora. 1984. Una tabla de doble entrada para medir la biomasa de lena y forraje en condiciones de la Finca La Francia, Siquirres, Costa Rica. CATIE, Turrialba, Costa Rica, 8 p. (mimeo).
5. Bond, W. E. T. 1944. Hedge plants in Northern Nigeria. Trop Agric (Trinidad) 21:228-230.
6. Budowski, G. 1982. The socio-economic effects of forest management on lives of people living in the area: the case of Central American and some Caribbean countries. In: E. G. Hallsworth (ed), Socio-economic effects and constraints in tropical forest management, pp 87-102. John Wiley, New York, NY.

7. Budowski, G. 1987. Fence posts in Costa Rica. United Nations Univ, Tokyo (in press).

8. Budowski, G., Russo, R. O. and E. Mora. 1985. Productividad de una cerca viva de _Erythrina berteroana_ Urban en Turrialba, Costa Rica. Turrialba 35:83-86.

9. Burgos, J. A. 1952. Postes vivos para cercos. Circular extension no 39. Estacion Experimental Agricola, Tingo Maria, Peru. 6 p.

10. CATIE. 1986. Silvicultural de especies promisorias para produccion de lena en America Central: Resultado de cinco anos de investigacion. Centro Agronomico de Investigacion y Ensenanza (CATIE), Turrialba, Costa Rica. 228 p.

11. Combe, J. and G. Budwoski. 1979. Classification of agroforestry techniques: a literature review. _In_: G. de las Salas (ed), Workshop Agrofor Systems Latin Amer, pp 17-47. United Nations Univ and CATIE, Turrialba, Costa Rica.

12. Crane, J. C. 1945. Living fence posts in Cuba. Agric Americas 5:34-38.

13. Howes, F. N. 1946. Fence and barrier plants in warm climates. Kew Bull Misc Infor 2:51-87.

14. Lozano, O. R. 1962. Postes vivos para cercos. MSc Thesis, CATIE/IICA, Turrialba, Costa Rica. 77 p.

15. Otarola, T. A. and H. A. Martinez. 1985. Manejo y produccion de cercas vivas de _Gliricidia sepium_ en el Nordeste de Honduras. _In_: R. O. Russo (ed), Los arboles de uso multiple en sistemas agroforestales. Trabajos presentados en la Segunda Reunion del Grupo S1.07.07 - Agroforesteria. CATIE, Turrialba, Costa Rica.

16. OTS and CATIE. 1986. Sistemas agroforestales: Principios y aplicaciones en los tropicos. Organizacion de Estudios Tropicales (OTS) y Centro Agronomico Tropical de Investigacion y Ensenanza (CATIE). 818 p.

17. Picado, W. and R. Salazar. 1984. Produccion de biomasa y lena en cercas vivas de _Gliricidia sepium_ (Jacq.) Steud. de dos anos de edad en Costa Rica. Silvoenergia (Costa Rica) 1:1-4.

18. Sauer, J. D. 1979. Living fences in Costa Rican agriculture. Turrialba (Costa Rica) 29:255-261.

12. TRADITIONAL AGROFORESTRY PRACTICES OF NATIVE AND RIBERENO FARMERS IN THE LOWLAND PERUVIAN AMAZON

Christine Padoch and Wil de Jong

Institute of Economic Botany, The New York Botanical Garden, Bronx, New York 10458 and Instituto de Investigaciones de la Amazonia Peruana, Abelardo Quinones km 2, Iquitos, Peru.

12.1. ABSTRACT

Recent research on traditional agriculture in the Peruvian Amazon indicates that agroforestry practices are widespread and extremely varied. This article describes five agroforestry systems found in the vicinity of Iquitos, Peru. Although all begin as shifting cultivation fields, they differ greatly in species composition and richness, in intensity and length of management, in economic orientation, and in adaptation to particular ecological conditions. Four of the five systems are found not in tribal but in mestizo communities of the region. The information presented shows that basic traditional swidden-fallow agroforestry practices are adaptable to varying environmental and economic situations.

12.2. INTRODUCTION

In reviewing alternatives for development of both subsistence and commercial production of food and timber, as well as for restoration of degraded lands, the National Research Council (U.S.A.) called for increased research into traditional[1] agroforestry systems as possible models (11). As that call is heeded and more research on traditional patterns is done, it is becoming evident that many widely differing systems are used by traditional peoples. Perhaps the most important of these in terms of economic potential and wide distribution throughout the humid tropics are those systems based on succession in swidden (shifting cultivation) fields. Recognition that swiddeners tend also to be agroforesters is not altogether new. Researchers have often reported that shifting cultivators in Asia, Africa, and the Americas plant tree species within their fields. However, in the past, little attention was focused on the management and importance of perennials in most swidden systems. Changes in species composition in the later phases of swidden field succession, methods of cultivating or protecting planted or spontaneously occurring trees in swidden fallow fields, and variation in quantity and type of product harvested from older fields have tended to go uninvestigated and unmentioned.

Such an oversight can largely be explained by the prevailing understanding of shifting cultivation as a system that includes a brief cropping period and a longer period of abandonment. The fallowing or temporary abandonment of fields has long been considered essential, but

---

[1]By "traditional" we do not mean to imply that these systems are unchanged since before European contact, or are unchanging. We merely mean that these systems use local products and local techniques.

its importance was previously considered strictly ecological, not economic. For the past three decades or more, research reports have documented that during this period of "rest" the swidden field recuperates its fertility and usefulness for further cropping. It was previously widely accepted however, that as a source of economic products, fallow fields contain little of importance.

The brief duration of most anthropological and geographical studies of shifting cultivation also accounts for the failure of many researchers to remark on the importance of swidden-fallow management and use. The majority of investigators have followed swidden field use for no more than a year or two. Fallow fields, however, tend to be used for many years afterwards, and their management, often minimal and quite informal, is frequently difficult to detect.

Reports on shifting cultivation systems which place due emphasis on the swidden fallow field as a continuing, albeit changing, source of food, fiber, medicines, and other economic products are now coming in from various parts of the world. Some pioneering studies, such as Gordon's description of swidden fallow management and use along the Caribbean coast of Panama (6, 7), are being newly appreciated. And many of the classics of the shifting cultivation literature, including the works of Conklin, Spencer, Clarke, Denevan, and Harris (2, 14, 1, 3, 8), are now being reread and references to cultivation of perennials and other indications of fallow use are being noted.

Apart from the references given above, swidden-fallow use and/or the planting of tree crops in swidden fields is mentioned in surprisingly few Amazonian ethnographies or other descriptive works. (For a summary of these references see Ref. 5:346). However, as the concept of traditional swidden-fallow agroforestry becomes better known and researchers throughout the Neotropics learn to look for tree-enriched fallows, it is becoming clear that such practices are the norm rather than the exception in traditional Amazonian agriculture. In our research in the lowland Peruvian Amazon, we have found many examples of such resource use in both tribal communities and ribereno[2] villages.

Not only is economic use of fallows virtually ubiquitous in the Peruvian Amazon, it is also tremendously varied. Within a radius of 150 kilometers of the city of Iquitos, in a sample of only four villages, we have observed agroforestry systems differing significantly in the species included, in the management practices employed, and in economic production and income realized. These different configurations can be identified as responses to varying ecological and economic conditions, as well as to the differing needs, abilities, and tastes of farming households.

In this article we shall describe five examples of traditional systems with which we are especially familiar. These examples share some characteristics. All began as swidden fields; then perennials gradually replaced annuals as the predominant crops. However, major differences are also notable, particularly in species richness and management patterns. Several of the fields described should perhaps not even be termed fallows. For instance, we shall describe systems where management is both intensive

---

[2] In the lowland Peruvian Amazon, all farmers who do not reside in tribally organized communities are known as riberenos, mestizos, or campesinos. Some persons classified as riberenos may in fact speak an indigenous language and have begun their lives as, officially, natives. Many are of mixed Amazonian and European ancestry.

and continued; the resulting fields might better be called permanent orchards. However, all the examples represent dynamic traditional agroforestry systems which change in composition and management requirements over time.

The few types that we shall present should be considered neither exhaustive of the variations found, nor necessarily the most typical of agroforestry systems of the Iquitos area. After presenting and comparing these examples of traditional resource use practices, we shall attempt to present a number of suggestions for further research in the area, and point out several aspects of these systems which may serve in planning for enhanced agricultural production in the humid tropics.

12.3. SWIDDEN FALLOW AGROFORESTRY AMONG THE BORA INDIANS.

Most of the literature that mentions swidden-fallow use in Amazonia -- indeed most information on any type of traditional resource use in the region -- is limited to studies of tribal groups. Perhaps the most complete project to date devoted specifically to the study of swidden-fallow agroforestry among an indigenous group of Amazonia, was a joint effort of researchers from the University of Wisconsin (U.S.A.) and the Universidad Nacional de la Amazonia Peruana. The project focused on use and maintenance of swidden fallow fields by the Bora Indians of the Yaguasyacu river in northeastern Peru. The investigators, who included geographers, anthropologists, foresters, botanists, and agronomists, examined and described the major features of Bora fallow use and looked into the changes that occur in swidden fields with increasing age. (For more detail on Bora agroforestry, see 4, 5).

The center of the study, the Bora community of Brillo Nuevo is located along the Yaguasyacu River, a tributary of the Ampiyacu River, which itself joins the Amazon around the town of Pebas, 120 kilometers northeast of Iquitos (see Fig. 1). In 1981 the village comprised 43 families, all descendants of tribal Bora who had been brought from neighboring Colombia in 1934. Although the people of Brillo Nuevo use many market items and have generally become quite assimilated into Peruvian society, we believe their subsistence activities still largely reflect traditional patterns. The distance of the village from Iquitos, the only large market of the area, and its location on a stream which in low water becomes virtually impassable, makes frequent market participation difficult for most of the residents of Brillo Nuevo.

The climax vegetation of the zone is Humid Tropical Forest. In the immediate environs of Brillo Nuevo considerable secondary forest (i.e., swidden fallow fields) can be found; however, mature forest remains close by the village. As in most Amazonian _terra firme_ sites, the most important agricultural cycle in Brillo Nuevo begins with the cutting, drying, and burning of new fields. Fields are made in both primary and secondary forest.

During the cutting and burning of vegetation, particularly valuable species, including fruit trees, palms used for fiber, and important timber species tend to be spared. The plots are then planted with a variety of annual, semiperennial and perennial crops; manioc (_Manihot esculenta_), the dietary staple, predominates in these new fields. (See 4:350-351 for a list of common Bora corps.) The number of crop species in some fields is very low, with cultigens almost exclusively restricted to annuals. In contrast, other plots are planted to a great mix of species including a variety of perennials. Often included are ten or more species of fruit

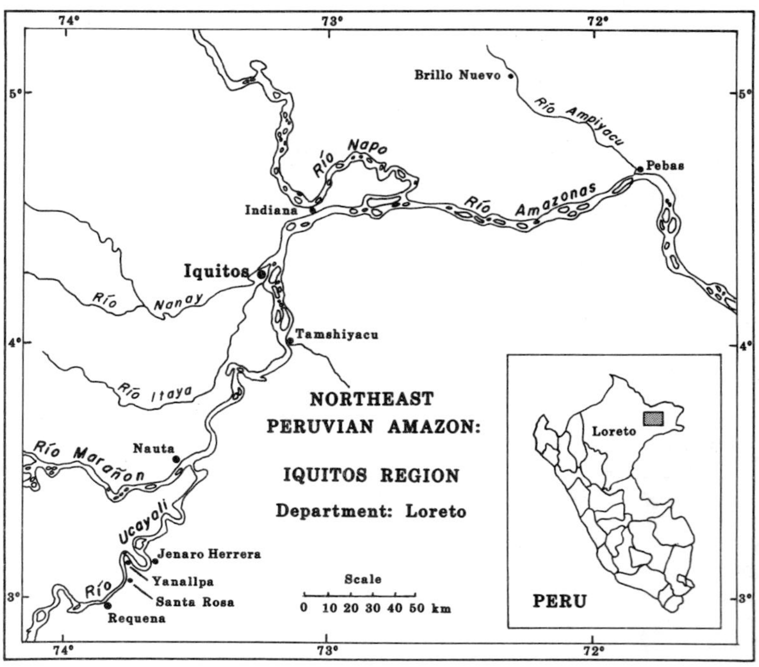

Figure 1. Location map for the five agroforestry study areas described in this paper.

trees as well as a number of other woody plants such as coca (<u>Erythroxylon coca</u>) and barbasco (<u>Lonchocarpus sp.</u>), a fish poison.

Cropping with manioc usually is continued for two consecutive years. Subsequent to these plantings, the less actively managed phase, or transition from field to fallow begins. In the meantime, the Bora cultivator clears other sites to plant his or her annual crops. Thus a farmer always has a large number of producing plots: some in early phases of swiddening, others passing into the fallow stage, and still others in the mature forest-fallow phase.

The assemblage of plant species found in a Bora swidden-fallow plot of any given age, depends on a great many variables, only one of which is the list of crops originally planted. We have mentioned that in initial forest clearing important fruit and other trees tend to be protected from cutting and burning. These individuals can subsequently be found in the fallow field. Other spontaneously occurring plants resulting either from the coppicing of cut vegetation of the sprouting of seeds are also found. If such plants are used for some purpose by Bora villagers, even several weedings may leave them untouched. In fact, weeding may be done with the express purpose of aiding the growth of valuable "weed" species. Which original crops survive also depends to a degree on the care given to them.

182

Many of the individual plants of a species that was originally planted may actually be volunteers that result from natural seeding.

Thus, from a variety of sources and resulting from different processes, a Bora fallow field, 5 years after original planting, may comprise a large variety of useful plants. One such field that was mapped during the research project still contained a recognizable manioc patch, but the plants were small and produced little (4:350-352). Six of the 12 other original cultigens were producing well and were harvested by the field's owners; among these figured four fruit species, coca, and barbasco.

Natural vegetation had encroached considerably on the outer edges of the original field, since weeding was restricted largely to areas immediately around important economic plants. The naturally regenerating area contained a great number of immediately or potentially useful species. For instance, materials for basket-weaving and other utilitarian purposes could already be harvested from the young vegetation, and other species promised use, when mature, as construction woods.

Researchers found that, commonly, some weeding and other maintenance work in swidden fallows continued for about twelve years after planting. Older fallows mapped and analyzed by the project revealed that an abundance of useful species remained in old fallows well through the first decade after original clearing. For instance, a 9-year-old fallow sampled at Brillo Nuevo still contained five species of cultivated fruit trees as well as coca, although all but the coca were represented by few individuals. In addition, numerous useful protected but not planted trees were found, including such valuable construction timbers as tropical cedar (<u>Cedrela</u> <u>odorata</u>).

From the very beginnings of the fallow stage, and continuing throughout the cycle, swidden-fallows present a very patchy appearance. While total management activity in the plot diminishes, it should be noted that this decrease in weeding is not even throughout the field. Some limited areas are actively cleaned for many years, while other areas are left virtually untouched. The cultivated area shrinks in size as the forest encroaches on the original cleared area.

In older fields, human management is minimal and apparently more haphazard. A farmer visiting such a site to collect fruits or other useful materials, may still clear a few lianas or other plants which threaten to interfere with an economically valuable species. But even after two decades of growth, some cultigens are encountered on old swidden sites and their products are harvested. In a field that was sampled 20 years after original clearing, researchers still encountered three clearly cultivated fruit species, and another 53 useful (but not planted) species. Included among the latter were construction materials, medicines, and foods; plants used for basket-making, for salt-making, and for extraction of pitch were also found in quantity.

In even older fields, 30 years and more in age, a few cultivated fruit trees have been observed to survive, but the majority of useful trees at this stage fall into the spontaneously occurring or protected category. Construction woods, medicinals, and handicraft materials are some of the valuable products that occur in these older stands.

Another important resource of swidden-fallow fields is game (see also Eisenberg and Harris, this volume). Attracted by secondary vegetation artificially enriched with planted and protected fruit trees, in certain seasons animals tend to be encountered in fallow fields in greater concentrations than in the forest. Villagers routinely build blinds near

fruit trees to await and hunt down birds and mammals, particularly monkeys
and large rodents (10).

We wish to stress that the description of Bora fallow use that we
have very briefly sketched here fails to convey the full range of
variability found in Brille Nuevo fallows. We mentioned some of the
sources of variation in the species compositon of fallow fields. However,
numerous geographical, cultural, ecological, personal, and purely chance
factors also intervene in determining the exact composition of a swidden
fallow. The following variables may be mentioned as a partial list: the
age of the vegetation originally cut and its previous use, the location of
the field in reference to the farmer's residence, the composition of other
fields that the owner already has in production, the particular conditions
in soil, drainage, slope, and aspect of the site, and the personal
preferences of the agriculturist.

Although variable in species composition, most fields, once they have
passed into a fallow stage, are rich in useful plants. The majority of
fields develop from swiddens planted with diverse cultigens, and
subsequent invasion by secondary vegetation makes them even richer in
plant species. Project researchers identified 133 species of useful
plants within the fallow fields sampled. Agroforestry systems such as
those of the Bora, may be seen as particularly useful for agriculturists
whose principal goal is to satisfy the diverse daily needs of their
households; they need to produce a little of a lot of different things.
Large plantations of any single species would be of little use in a small
community with difficult access to a larger consumer market. However, not
all Amazonian agroforesters manage their fields for high diversity of
useful species. The pattern which broadly characterizes Bora fallows
should not be considered a ubiquitous or necessary feature of Amazonian
agroforestry systems.

12.4. MARKET-ORIENTED AGROFORESTRY IN TAMSHIYACU

Upriver and approximately 30 kilometers southeast of Iquitos, stands
the town of Tamshiyacu. The town is noted as the most important native
fruit producing area for the urban market. Its principal products are
pineapples and umari (_Poraqueiba sericea_) fruit. Both of these cash crops
are cultivated within a swidden-fallow agroforestry system. Pineapples
are produced during an early stage of the cyclic system; umari is the
principal product of the late stage. Unlike the subsistence-oriented
agriculture of Brillo Nuevo, a market emphasis pervades virtually every
phase of this particular agroforestry system. For more detailed
information on Tamshiyacu agriculture see (9, 12).

A brief investigation was carried out in Tamshiyacu as part of the
Bora research project. Further research was done under the auspices of
the research program of the Peruvian Ministry of Agriculture's Instituto
Nacional de Investigacion y Promocion Agropecuaria (INIPA).

The people of Tamshiyacu are not officially recognized as Indians
native to the region, although few if any Tamshiyaquinos are recent
arrivals. They are members of the majority population of the Peruvian
Amazon, the riberenos. These descendants of detribalized natives and
immigrants who mostly arrived during the great rubber boom of the turn of
the century, are a population until now curiously invisible to both
researchers and the government outside of the Amazon, both of whom usually
confuse riberenos with recent colonists.

The town is one of the larger ones in the region, with a population of approximately 2000 persons. Although the majority gain most of their livelihood from farming, many Tamshiyaqinos maintain frequent contact with Iquitos and its markets. "Collectivos" (river buses) bound for the city pass three times a day. Two local boats make the trip before dawn which allows villagers with products to sell to reach the urban market during its most active hours.

In climate and forest type Tamshiyacu is not unlike Brillo Nuevo. Mature forest however, is to be found at greater distance from this larger, longer-settled town. Like the Bora village, Tamshiyacu and its agricultural fields are located on high ground, far above the reach of the Amazon's flood.

Despite their official non-native status, Tamshiyaquinos are clearly inheritors as well as adapters of indigenous resource use traditions. Their agricultural cycle begins with the clearing of forest, both primary and secondary. Again, any valuable forest species of fruit trees left from earlier cropping cycles that are encountered will be spared. However, unlike in Brillo Nuevo and among most other shifting cultivators, in Tamshiyacu all cut and slashed vegetation is not burned. Much of the dried wood is gathered and transformed into charcoal for sale in Iquitos.

Many of the crops planted in Tamshiyacu swiddens are identical to those found in Bora plots. Frequently the choice and quantity of cultigens included in a field are, however, determined less by household subsistence needs than by recent and expected urban market price trends. In the initial swidden phase, manioc and plantains tend to be the most important cultigens and are replanted after the initial harvest. These, like all subsequent crops are destined for both household comsumption and market sale. A variety of other annual crops are intermixed in the first planting, pineapples among them. However, in the second year, a number of perennials are also included. Manioc declines in importance after the second year, but pineapples, a major cash crop, continue to be harvested through the fourth year after clearing. By that time or even earlier, however, yields of most of the original annuals have declined or ceased altogether, and early maturing fruit trees are the major producers.

Among the trees that are commonly planted in Tamshiyacu fields are fast-growing but shortlived species such as cashew (<u>Anacardium occidentale</u>), uvilla (<u>Pourouma cecropiaefolia</u>), and caimito (<u>Pouteria caimito</u>). These trees bear fruits that have a ready market in Iquitos, but their yield tends to decline or harvesting becomes difficult after five or six years. However, in the fourth or fifth year since swidden clearing, umari (<u>Poraqueiba sericea</u>), the fruit for which Tamshiyacu is most famous, begins to yield in quantity. Continuing for twenty or more years, the fruit of the umari tree is the most important economic product of the swidden fallow or orchard.

Actually, umari and, in a far smaller quantity, Brazil nuts (<u>Bertholletia excelsa</u>), are virtually the only products of the later stages of the swidden cycle. (Brazil nut trees are often planted in small quantities.) Both of these fruits are largely taken to Iquitos' markets for sale. In its prime producing stage (i.e., from five to twenty years after planting), a one hectare field of umari will annually yield up to 80,000 fruits, a quantity far above a household's ability to consume. Many Tamshiyacu households own several hectares of "umaral" or umari orchard.

Maintenance of the swidden and of the umari fallow-orchard declines in intensity, as in the Bora system, with the field's age. However, in

contrast to Brillo Nuevo's swiddeners, the farmers of Tamshiyacu continue
to clean their fields quite thoroughly throughout the entire fruit
production phase, which may continue for 25 or more years. Frequency of
cleaning is reduced from several times a year during the stage when annual
crops are in production, to once or twice a year at the time of the
harvest of umari. The care with which cleaning is done also changes.
Careful weeding gives way to more rapid slash weeding as annual herbaceous
plants are replaced by tree species. However, the secondary forest is
never allowed to reestablish itself in a Tamshiyacu agroforestry plot
while umari is still in production. Although a very few extraneous
naturally occurring fruit or timber trees may be protected, all other
vegetation is cut. Such continued clearing not only keeps competition
down for the umari crop, it is also essential for the harvest of the
fruit. Umari fruits cannot be plucked from the tree; they are gathered
when they fall to the ground.

The result of this active and intensive maintenance is, of course,
much higher fruit production in Tamshiyacu fields, and much lower species
diversity of both economically useful and other plants, than in Brillo
Nuevo. In the two settlements, farmers, faced with different needs and
presented with different opportunities, manage their swidden fallows with
very different goals in mind.

As soon as umari yield begins to decline significantly, the umari and
other large trees present in the plot are cut down and again made into
charcoal. (Umari is said to produce exceptionally good charcoal.) Brazil
nut trees, however, are often spared.

Following this second charcoal-making, the field is generally
fallowed for a period of five to ten years. During this time, secondary
vegetation is allowed to invade the plot and no significant management is
carried out, although some collection of minor forest products certainly
occurs. These few years of natural succession, allege village farmers,
are sufficient to restore the field to a condition appropriate for another
swidden-orchard production cycle to begin. Some umarales never pass
through this fallow phase; immediately after cutting and burning, annual
crops are planted.

The swidden fields of Tamshiyacu again show much greater variation
than this brief sketch suggests (see also 9). For instance, some farmers
burn their fields without making charcoal. Most agriculturists make a few
fields where perennials are never included; after two harvests of manioc
and several of pineapple and other cultigens, these fields revert to
unmanaged natural vegetation. In addition, it should be noted that the
residents of Tamshiyacu, like most rural dwellers of the Peruvian Amazon,
engage in various other subsistence and cash producing activities. Most
farmers are also hunters, fishermen, occasional wage-laborers or even
market vendors.

In a brief economic survey that was carried out in the town in 1983,
we found however, that the sale of cultivated fruits was the chief source
of income of village residents (13). With the exception of a small
quantity produced in house gardens, virtually all marketed fruits were
products of the cyclic swidden orchard/fallow system described above. The
incomes earned from the sale of their agroforestry products varied
enormously and our small survey will not allow us to cite any reliable
figures as averages. We can state with confidence, however, that owners
of extensive umarales did gain several times the cash incomes of average
Amazonian farmers. Household incomes of approximately U.S. $5,000 were

enjoyed by several active umari producers that we interviewed. Few rural
residents of the region ever earn half that income annually.

The Tamshiyacu system outlined here bears considerable resemblance to
the agricultural practices of the Bora, and can be considered a
transformation of the indigenous pattern. In Tamshiyacu, production in
later phases of the cycle is highly specialized, reflecting the economic
opportunities offerred by a large nearby market. The pattern typical of
Brillo Nuevo, on the other hand, is far more generalized and diverse and
thus adapted to satisfying the varied needs of the subsistence farmer and
the limited consumption capacity of a relatively isolated small
settlement.

The specialization of the Tamshiyacu system obviously has its
monetary rewards, but it also has its pitfalls. Umari production is
seasonal and even Iquitos with its quarter-million population offers, in
years of particularly high production, insufficient demand to absorb all
the fruit that Tamshiyacu produces. To date, no suitable process for
preserving the bland, oily pulp of the umari fruit has been developed and
all fruit is consumed fresh. However, those farmers who have several
fields in distinct phases of the agroforestry cycle can enjoy a reasonable
income over at least eight months of the year. The town's two principal
cash crops, pineapples and umari, tend to yield in different months,
staggering both labor demands and income.

Another obvious characteristic that distinguishes the two examples is
the relatively intensive control of natural succession in Tamshiyacu
fields. In the case of the mestizo town, old swiddens are maintained as
orchards for a long time before they are allowed to become true forest
fallows. In Brillo Nuevo the process is more gradual.

12.5. THE DIVERSE ORCHARD-FALLOWS OF SANTA ROSA

Our third example combines aspects of the two systems thus far
described. This type of agroforestry field is doubtless to be encountered
in many villages along the Amazon and its major tributaries in the Iquitos
region. We are studying it in the village of Santa Rosa along the lower
Ucayali River, about 150 kilometers upstream from Iquitos. This research
is being done as part of a cooperative program between the Instituto de
Investigaciones de la Amazonia Peruana, and the New York Botanical
Garden's Institute for Economic Botany.

Santa Rosa in 1986 had a population of slightly over 300 persons.
Although officially designated mestizos (non-Indians), the community is
actually a very complex mix of representatives or descendants of several
tribal groups (now acculturated), and of families of some European
background but with a long history in the region. Many Santa Rosinos
recount involved personal histories including several long economic
migrations and changes of residence.

The village is rather new. Many present residents once worked on a
large plantation just a kilometer or two upriver. That enterprise folded
about thirty years ago and the community of Santa Rosa was founded several
years later. The location of the village itself has moved a number of
times in the last few decades. The present site is on <u>terra firme</u>,
although villagers also farm many lower floodplain areas, including
"barreales" and "playas" (mud-flats and beaches that flood annually), and
"restingas" (natural levees that flood periodically). The instability of
residence is largely an effect of their living on the banks of the
ever-meandering Ucayali. The river has already swallowed several of their

old village sites. Moves made by individual families are occasioned by diverse motives. A common one is the search for secondary schooling for their adolescent children. Requena, a town of approximately 10,000 inhabitants and a regional educational center, is only 20 kilometers up the Ucayali. The town also offers a convenient, if rather limited, market for Santa Rosa's produce.

Agriculturally, the village differs from Brillo Nuevo and Tamshiyacu, most notably because residents engage in floodplain, or barreal and restinga, farming. Most Santa Rosa farmers spend less than half their working time cultivating the terra firme swidden plots that we shall describe here. However, at any time, almost everyone has at least one producing swidden in a terra firme site, and many maintain agroforestry plots. These are characterized by a high number and diversity of cultivated trees and by a pattern of long-term continued maintenance.

All Santa Rosa agroforestry fields are established in largely the same way as in Brillo Nuevo and Tamshiyacu: swiddens are cut and cleared. The practice of transforming some cut vegetation to charcoal, rather than burning it all in the field, is not unknown but is rare. The assemblage of annuals and semiperennials first planted into the swidden is similar to that found in Tamshiyacu: a combination of subsistence necessities and some crops destined for the market. In many swiddens, cropping and maintenance cease after two or three years of intensive annual crop production. These fields are then fallowed, often for a brief period of five to six years, and then used again. However, most farmers choose to turn at least one or two of their swidden fields into agroforestry plots. When such a choice has been made, planting of fruit trees usually commences in the first year of swidden production, with more and more tree species planted as the field ages.

Some of the early fruiting trees will be harvestable in the second or third year after original clearing. By about the fifth or sixth year the swidden has become an orchard. The orchard fallows characteristic of Santa Rosa differ significantly from the Tamshiyacu example. While the latter are overwhelming producers of one species, umari, Santa Rosa orchards yield a large array of fruits and may be said to more closely resemble the fallows of the Bora. However, the similarity is slight; Santa Rosa's orchards contain far more species of planted trees while Brillo Nuevo residents harvest many of their products from spontaneously occurring species. This contrast partly reflects different planting strategies but in large measure it results from the continuing and relatively intensive management that Santa Rosinos practice in the their fallows.

As an example of the Santa Rosa system, we can cite the field of Don E., who differs from much of the population only in that he and his family have lived on and farmed the terra firme part of the settlement longer than many of their neighbors. The plot was first cleared by Don E.'s father approximately 25 years ago, and a swidden plot was established. Although we have little precise information on the early stages of this plot's cycle, many of the larger trees now in the field were planted within the first few years after clearing. For instance, several mangos (Mangifera indica) are impressively large with diameters of 75 and 54 cm. These figured among the orchard's original plantings. Mature examples of zapote (Quararibea cordata) and of mamey or pomarosa (Syzygium malaccense) also testify to the orchard's age. Apart from fruit trees, large timbers such as tropical cedar (Cedrela odorata) -- in this case planted by Don E.'s father -- are found in the plot.

188

However, many of the fruit and other cultivated tree species in Don E.'s orchard were not planted two decades ago. Throughout the field's existence, the present owner and his father have been planting species which they considered useful. Thus the field contains trees and even some herbaceous plants of all ages, and the diversity of crop plants is very high. Among the unusual plants we encountered in Don E.'s field are several useful palms which are very rarely planted, including cashapona (_Iriartea exorrhiza_), shebon (_Socratea_ _sp._), and yarina (_Phytelephas macrocarpa_).

Most of the plot has an open appearance, resulting from the continuing maintenance work performed. Don E. clears his plot by slashing all the secondary vegetation once or twice per year. The only new plants that he may spare are volunteers that grow from seeds of already established cultigens. Several tropical cedars and some of the individual fruit trees in Don E.'s field should not strictly be considered planted, as they result from such natural seeding. However, little or no forest regrowth is allowed to survive the field's cleanings. We should note that in one other intensively managed orchard in Santa Rosa, we did find some preservation and protection of forest species. The growth of individual pichirina (_Vismia_ _sp._) and rifari (_Miconia_ _sp._) trees was encouraged in areas where fruits had not yet been planted. Awaiting the availability of appropriate fruit tree seeds or seedlings, the farmer protected these secondary tree species since they helped prevent excessive weed invasion of his orchard. These trees may later be used for house construction. Many species of naturally occurring fruits, fibers, medicinals and other useful forest products are gathered and used by Santa Rosinos. However, these tend to be sought in the non-managed fallows that abound, and in the primary forest which remains close by.

Exactly how long a field of the type maintained by Don E. will continue to be used and cleared is unknown to us. As we mentioned above, the village itself is young and most residents have been farming the _terra firme_ areas for less than two decades. Some fields, such as Don E.'s, have been continuously cleaned since they were made. Several have been abandoned to second growth because their owners left the village for life in neighboring Requena or in another village. The high degree of mobility that characterizes the mestizo population of the Peruvian Amazon is an important factor in explaining orchard-fallow abandonment.

Besides its age, Don E.'s plot may be considered somewhat unusual in the village because of its extreme variability. Don E. is a man who likes plants, likes to see variety in his garden, will go far out of his way to get a novel plant, and prides himself on the impressive garden he manages. In this interest and pride, he may be somewhat atypical; not every farmer in Santa Rosa would choose to plant palms in his garden that might be available in the neighboring forest. However, in our visits to many villages of the region we have frequently met individuals of Don E.'s type. We believe that such exceptional people perform an important role in their particular communities, serving as maintainers of seed banks. Every village probably has one such person. Santa Rosa has several.

The products of the diverse orchards of Santa Rosa serve both the subsistence needs of their owners and families and provide products to sell in the market. In contrast to their counterparts in Tamshiyacu, the agroforesters of this village produce very little of any particular fruit. On the other hand, their fields do yield a variety throughout the year. We mentioned above that the substantial market of Iquitos often cannot absorb the production that a large umari harvest will offer. Transport of fruit from Santa Rosa to Iquitos is difficult, since cargo and passenger boats travel only twice a week, and fares and cargo charges are high for local residents. Requena is the usual market for Santa Rosinos. Given the small population of Requena, production of a small quantity of many fruits, rather than specialization in one, is economically advantageous.

It should also be noted that it is the lowlying areas, rather than the poorer soils of the <u>terra firme</u> that produce Santa Rosa's cash crops. Rice cultivated on seasonally inundated mud-flats and plaintains produced on the natural levees yield most of the cash that is earned.

As we noted in Brillo Nuevo, the farmers of Santa Rosa will often include a few valued timber trees in their orchards. We have met only a few people in our sample communities who have sold the tropical cedars that they have planted. However, many people consider their planted or protected timber trees as a reserve, as money in the bank.

12.6. VARZEA AGROFORESTRY IN YANALLPA

As our last example we shall describe the practices characteristic of the settlement where we are now engaged in research, the varzea village of Yanallpa. Our work in this community is a continuation of the project begun in Santa Rosa. Since the study, which is to extend over a year, has only recently begun, the information presented here will necessarily be rather cursory. However, Yanallpa bears mention since, in contrast to the other villages cited, this community only five kilometers downriver from Santa Rosa, lies entirely within the floodplain of the Ucayali River. (At the height of the exceptionally strong 1986 flood, all parts of the village, including all its agricultural fields, were inundated to a depth of at least a meter). The floodplains of Amazonian white-water, sediment-laden rivers have long been recognized as the region's most fertile areas and thus most suited for intensive cultivation of annuals. Agroforestry has only recently been considered as a production alternative for such environments.

Yanallpa has about the same number of inhabitants as Santa Rosa, but has a much longer history as a village. For about a century, small farmers and forest produce gatherers have made the high natural levee along which the village is situated, their home. In contrast to Santa Rosa, Yanallpa was never the site of large plantations. It has long been considered a fine place for subsistence and market farming. The soils of the restinga produce crops of plantains, corn, and papayas in quantity. The farmer however, must run the risk of losing all those crops periodically, when an exceptionally high annual flood appears. Long term residents insist that in past decades such floods never covered the whole village; only the lower parts of the levee were inundated. The entire restinga has flooded three times within the last seven years, in 1979, 1984, and in 1986.

Apart from the ten kilometer long natural levee along which the village stretches, Yanallpa's farmers also exploit a low area which borders on a backswamp, an extensive beach which has recently formed in front of the village, and some mud-flats directly across the Ucayali. On the lowest parts which are flooded every year, rice, cowpeas, and beans are raised.

Corn, plantains, manioc, and assorted fruits are planted on the higher levee. Most fruits are produced in multi-species orchards that share some of the characteristics of the orchards we described in Santa Rosa. In this case they are not fallows at all, but permanent plots of fruit trees interspersed, in many instances, with timber and other species. These orchards also are very diverse, but only include species that can survive occasional inundation. In one of the gardens that we censused, we found a total of 29 useful species, 20 of them fruit trees, 2 timber species, a palm used for thatch, one rubber tree, and five herbaceous plants. Three of the herbaceous species would not have survived the flood, had they not been dug up before the waters reached the garden and kept in receptacles until the flood receded. Yanallpa farmers insist that prior to the recent very high floods, their gardens comprised many more species including such fruits as caimito (<u>Pouteria</u> <u>caimito</u>), zapote (<u>Quararibea</u> <u>cordata</u>), taperiba (<u>Spondias</u> <u>dulcis</u>), and a variety of citrus species. The three recent inundations have forced Yanallpa residents to change the species composition of their gardens, adapting themselves to periodic floods. Some residents are actively seeking flood-resistant plants to include in their levee gardens. The owner of one of Yanallpa's most diverse gardens, an older woman, Dona T., has transplanted several plants from the forest that she erroneously identified as desirable fruits. The inclusion of these non-producing trees in her garden testifies to the experimental spirit of the farmers of Yanallpa and other traditional communities.

The levee orchards largely yield fruits and other goods for household consumption. A part of the fruits harvested are marketed, usually in the town of Requena. Flood-resistant citrus varieties, like the toronja, a lime-like citrus, are among the most important commercial fruits. Many of the species that the recent floods killed, such as the caimitos and zapotes, previously were important money-makers.

The exact age of the orchards is impossible to determine, since management and replanting of fruit and other species is a continuous process. We assume that the original clearing and planting of the sites involved a swidden process similar to the other examples we have given. However, many of today's orchards have clearly existed for several decades and therefore Yanallpa's orchards cannot strictly be considered swidden-fallow plots.

Another agroforestry system worth noting in Yanallpa involves the production of annual and semi-perennial crops (plantains) beneath a broken canopy of varzea tropical cedar trees ("cedro blanco", <u>Cedrela</u> <u>sp</u>.) and other timber and fruit trees. This is a very common practice in Yanallpa. When a new field is made in a site where tropical cedars grow naturally, they are usually left in place, and short-cycle crops are annually planted beneath them. Other valuable spontaneously occurring plant species, including palms, are also often left standing. In some cases the cedros blancos and other trees do not occur naturally but are planted. The richer soils of the floodplain allow farmers to plant their annual crops for more than the one or two years possible for swiddeners. We have yet to determine how long annual cultivation of short-term crops can continue in any particular Yanallpa field before it must be fallowed. In recent years cultivation has been frequently interrupted by flooding which restores the soils and eliminates pests as would a swidden fallow.

We have seen half-hectare fields that include from four to twenty or more varzea cedars and other timber trees. The number of large trees left standing or planted depends on several factors. Among these is the choice of crops the farmer wishes to raise; yuca, for example often does poorly under shade, while plantains appear to thrive. Maize fields typically will include only a few large trees. The age of the particular plot may be a determining factor as well. In some instances, farmers plant only a few cedros blancos in their fields. But with time and natural seeding these increase to a significant stand of timber trees. Farmers continually make decisions as to which seedlings to preserve and which to cut.

At least one Yanallpa farmer, Don A., has allowed the forestry species in his fields to shade out the annuals and semi-perennials. He now manages and maintains one plot close to his house for production of forest products for his own and his neighbors' house-building needs, as well as for sale to local sawmills.

12.7. CONCLUSIONS AND RECOMMENDATIONS

We have presented descriptions of agroforestry practices found in four communities in the Peruvian Amazon. Only one of our sample villages is inhabited by members of a tribal group, the rest are settlements of non-tribal, but nonetheless largely indigenous Amazonians, variously known as riberenos, mestizos, or campesinos. Although all the communities described lie within a 150-km radius of the city of Iquitos, the problems and opportunities that each local area presents to its residents differ considerably.

We have sketched the principal features of five agroforestry types that we investigated in the region. These systems share some characteristics: they all are or were initiated as rather similar swiddens, they all undergo changes in species composition, and the management practices in all five change with time. The differences among the systems are perhaps more important: some become more diverse through time, and others less. Some are managed intensively over many years, others not. Some are very obviously designed to meet the subsistence needs of a small household, others are oriented toward production for a large market. Some focus on fruit, others on wood production. Some are adapted to periodic flooding, others lie well above the flood's reach.

These descriptions illustrate our statement that agroforestry systems in the Peruvian Amazon are not only very widespread, but are also varied. There is not one Amazonian agroforestry system. What might be considered the basic pattern of swidden-fallow agroforestry is a highly flexible one, adaptable to various environmental and economic conditions.

The examples that we have presented should not be considered static systems. We have mentioned that active experimentation occurs in Yanallpa, as farmers attempt to accomodate their agroforestry practices to changing flood regimes. The same dynamism and experimentation may be found in other communities.

We have noted that the few descriptions given here are not an exhaustive summary of agroforestry types found in the lowland Peruvian Amazon. On the outskirts of Iquitos itself, we have observed other adaptations that we have not had the opportunity to study. For instance, immigrant farmers to the Iquitos-Nauta road now under construction, are already supplying the urban market with substantial quantities of fruits and other goods. These crops are produced in swiddens and orchards located on very poor, sandy soils. We believe that a broad and long-term research effort in the area would yield results of value not only to the Iquitos area, but possibly to other tropical regions plagued with similarly poor agricultural conditions. Opportunity and need for research on traditional agroforestry systems in the area continue to be great.

We would particularly like to emphasize the need to study traditional systems of non-tribal communities. The forgotten villages of the Peruvian

Amazon (and of Brazil (13)) often present far more useful models of resource use planners than do the remote, tribal communities that have been given more attention. Most riberenos or mestizos living within a day's travel of Iquitos, participate more actively in market trade than do natives on small isolated tributaries. Rather than using what may appear to be "purer" tribal systems as models, and speculating on how these might serve the Amazonian who is heading toward increased market involvement, ribereno patterns may offer a "ready-made" model.

Our investigations in the Iquitos area have pointed out the great importance of market access and market demand in determining the configuration of a particular agroforestry system. The examples taken from Tamshiyacu and elsewhere show that swiddeners can and do respond strongly to market opportunities; they can and do grow far more fruit than their household can consume. However, the market of Iquitos is limited and cannot consume even the yields that already are produced. In exceptionally heavy harvest seasons 30-50% of umari fruits produced in Tamshiyacu spoil on the ground. Research and extension in agroforestry systems must be accompanied by sound economic knowledge and advice. If the small farmers of the region are to be encouraged to augment their production of fruits and other agroforestry products, they must also be presented with opportunities for processing and extra-regional export of those products.

One of the continuing, major research needs is for in-depth and long-term research on local traditional agroforestry systems. Ethnohistorical information indicates that even some of the market-oriented types we have described have existed, with some modifications, for about a century. However, lasting effects on soils, changes in fallow structure and composition, and behavior of individual agroforestry species within mixed stands, have yet to be adequately investigated.

Finally, we wish to recommend that advice on both agricultural and forestry problems be made more available to traditional agroforesters. While there is often enthusiasm expressed in designing entire systems for small cultivators, little attention is given to helping present practitioners improve or refine their existing systems. For instance, we believe that with adequate advice on pruning and other forestry techniques, the quality of timber trees in Yanallpa and other Amazonian villages could be significantly improved and the incomes of the agroforesters further augmented.

ACKNOWLEDGMENTS

Since this article summarizes the results of several projects to which we have contributed we wish to thank a number of institutions and individuals for their aid. The Bora agroforestry project and the initial work in Tamshiyacu, were funded by the UNESCO Man and Biosphere Programme (MAB), under an agreement with the University of Wisconsin and the Universidad Nacional de la Amazonia Peruana. We would like to acknowledge the inspiration of Salvador Flores Paitan who pioneered agroforestry research in the Iquitos area, and William Denevan who directed the project. For their contributions in Tamshiyacu, we wish to thank Jomber Chota Inuma and John Unruh. Further research in Tamshiyacu done with funds from the Peruvian Ministry of Agriculture's Instituto Nacional de Investigacion y Promocion Agropecuaria (INIPA). Our special thanks go to Otoniel Mendoza, Director of the INIPA's field station in Iquitos and

Mario Pinedo, leader of the INIPA's Native Fruit Project in Iquitos. The Santa Rosa and Yanallpa research is a joint effort of the Instituto de Investigaciones de la Amazonia Peruana (IIAP), the New York Botanical Garden's Institute for Economic Botany. The Technical Cooperation Agency of the Swiss Government contributed some funding. Our special thanks go to Jose Lopez Parodi, director of IIAP's Centro de Investigaciones y Desarrollo, Jenaro Herrera, and director of the traditional agroforestry project. For their great patience and hospitality, we are especially grateful to the residents of Brillo Nuevo, Tamshiyacu, Santa Rosa and Yanallpa.
Figure 1 was drawn by Carol Gracie.

REFERENCES

1. Clarke, W. C. 1971. Place and people: An ecology of a New Guinean community. Univ Cal Press, Berkeley, CA.
2. Conklin, H. C. 1957. Hanunoo Agriculture. FAO, Rome.
3. Denevan, W. M. 1971. Campa subsistence in the Gran Pajonal, eastern Peru. Geogr Rev 61:496-518.
4. Denevan, W. M. Treacy, M., Alcorn, J. B., Padoch, C., Denslow, J. and S. Flores Paitan. 1984. Indigenous agroforestry in the Peruvian Amazon: Bora Indian management of swidden fallows. Interciencia 9:346-357.
5. Denevan, W. M. and C. Padoch (eds). (in press). Swidden fallow agroforestry in the Peruvian Amazon. Adv Econ Bot. New York Botanical Garden, New York, NY.
6. Gordon, B. L. 1969. Anthropogeography and rainforest ecology in Bocas del Toro province, Panama. Office Naval Res Rep, Dep Geog, Univ Cal, Berkeley, CA.
7. Gordon, B. L. 1982. A Panama forest and shore: natural history and Amerindian culture in Bocas del Toro. Boxwood Press, Pacific Grove, CA.
8. Harris, D. R. 1971. The ecology of swidden cultivation in the upper Orinoco rainforest, Venezuela. Geogr Rev 61:475-495.
9. Hiraoka, M. 1986. Zonation of mestizo riverine farming systems in northeastern Peru. Nat Geogr Res 2:354-371.
10. Linares, O. F. 1976. 'Garden hunting' in the American tropics. Human Ecol 4:331-349.
11. National Research Council. 1982. Ecological aspects of development in the humid tropics. Nat Acad Press, Washington, DC.
12. Padoch, C., Chota Inuma, J., Jong, W. de and J. Unruh. 1985. Amazonian agroforestry: a market-oriented system in Peru. Agrofor Syst 3:47-58.
13. Parker, E. P. (ed). 1985. The Amazon caboclo: historical and contemporary perspectives. Stud Third World Soc 32, Williamsburg, VA.
14. Spencer, J. E. 1966. Shifting cultivation in Southeastern Asia. Univ Cal Publ Geog, Vol 19. Univ Cal Press, Berkeley, CA.

13. AGROFORESTRY USING TAME PASTURES UNDER PLANTED PINES IN THE SOUTHEASTERN UNITED STATES

Clifford E. Lewis and Henry A. Pearson*

Range Scientist, USDA Forest Service, Southeastern Forest Experiment Station, Gainesville, FL 32611.

*Range Scientist, USDA Forest Service, Southern Forest Experiment Station, Pineville, LA 71360.

13.1. INTRODUCTION

Agroforestry is a term that has recently emerged in land management literature. Unfortunately, it means many things to many people because a variety of definitions have been used. Current application seems to encompass any land area that supports woody perennials which is also used for other agricultural purposes. However, including many of these land uses under "agroforestry" seems inappropriate.

A case in point is the inclusions of grazing natural forage plants under naturally occurring trees as a form of agroforestry. For example, Reid and Wilson (47) refer to rangeland forest grazing in native pine forest of North America as an extensive form of agroforestry. They include the pine-fir forests of the Pacific Northwest and Canada, pine forests in northern Arizona, and pine/pine-hardwood forests of the South. Batini et al. (2), while examining agroforestry practices in Australia, included the extensive grazing of indigenous forests. If all lands supporting woody plants--which are also grazed or cropped--are included in agroforestry, then the only land excluded would be prairies, tundra, and similar treeless areas.

Since the three major uses of land are for growing woody plants (primarily trees), herbaceous plants (primarily grasses) for livestock, and crops (including tame pasture), three forms of management can be identified as silvicultural, pastoral, agronomic. Areas of overlap become clear and can be logically named and defined. For this paper an important distinction is made between Silvo-pastoral, the grazing of natural forages that exist under naturally occurring forests (forest grazing) and Agro-silvo-pastoral, the combining of trees, forage crops (primarily tame pastures), and livestock.

If we can agree that agroforestry is "the deliberate growing of woody perennials on the same unit of land as agricultural crops and/or animals" (11), then "silvo-pastoral" as well as "agro-pastoral" are not agroforestry because silvo-pastoral is the grazing of natural forage under natural tree cover, and agro-pastoral does not include trees. This usage of the term agroforestry would help eliminate some conflicting terms currently in use (8, 35, 36, 37, 47). For these reasons, our discussions of agroforestry in the humid-subtropical zone of the United States will be limited to the deliberate combining of pines, pastures, and cattle.

13.2. HISTORY OF CATTLE IN THE SOUTH

Cattle were first introduced into the United Stated in 1520 or 1521 by Ponce de Leon on his second trip to Florida (9, 12). Subsequent attempts by Spain to explore and colonize this new world brought additional livestock into coastal Florida and other ports of the Gulf of Mexico. Most of the early attempts at colonization failed and Indian tribes adopted the abandoned or strayed livestock and became small-scale livestock producers. With the influx of European settlers from the northern colonies in the 1800's, additional cattle of European milk and beef breeds were brought into the South. These animals grazed primarily on native forages. Following the Civil War, cattle were sometimes grazed on small pastures of common bermudagrass (_Cynodon dactylon_ (L.) Persoon) and common carpetgrass (_Axonopus affinis_ Chase) that developed on abandoned cotton land.

With the development and introduction of additional and more productive pasture species, the use of woodland grazing began to decline. Coincident with this was the introduction and development of better beef breeds that required higher levels of nutrition than existed in native forages. Also at the same time, forestry became the dominant use of much of the South, especially with the development of pine planting techniques and plantation management in the 1920's. In attempts to develop management methods that would allow the production of both beef and wood, various techniques of integrated management were researched for using the native forages that existed in woodlands. However, interest in grazing in pine forests continued to decline since most animal scientists felt that better forage was needed for cattle than occurred naturally in the woods.

13.3. EARLY RESEARCH IN AGROFORESTRY

13.3.1. Exotic species under pines

The earliest attempts at agroforestry in the southeastern United States began in 1946 when 23 grasses and 14 legumes were tested over an 8-year period for their adaptability to southern forested conditions. Naturally-occurring stands of longleaf pine (_Pinus palustris_ Mill.) and slash pine (_P. elliottii_ Engelm.) were used to study the introduction of grasses and legumes without site preparation (soil disturbance), introduction of grasses and legumes with site preparation and fertilization, effects of various densities of pine canopy on grass/legume mixtures, and establishment of pastures on harvested forestland where 8 to 20 seed-trees/ha were retained.

In summarizing these studies (5, 16), it was determined that establishment of forage species was generally poor without mechanical site disturbance and that fertilization was needed to establish and maintain productive stands. A heavy tree canopy reduced yields and limited persistence of exotic forages. Best results were achieved with common carpetgrass, Pensacola bahiagrass (_Paspalum notatum_ Flugge), annual lespedeza (_Lespedeza striata_ (Thunberg) H. & A.) and white clover (_Trifolium repens_ L.). Pensacola bahiagrass was the most shade tolerant of the warm-season grasses.

13.3.2. Combining pines and pastures

The integration of pines, pastures, and cattle (also called intensive pine-pasture management) was studied over a 22-year period beginning in 1955 (28). Slash pine was planted in 1957 at spacings of 6.1 x 6.1 and 3.7 x 3.7 m and kept weed-free by mechanical cultivation for 3 years until trees reached a cow-resistant size. In order to compare tree growth, slash pine was also planted among natural understory vegetation and was neither cultivated nor fertilized. In the fourth year, pastures were planted with Pensacola bahiagrass, Coastal bermudagrass, and dallisgrass (_Paspalum dilatatum_ Poir.) under the pines and in open pastures with no pines. Annual grazing during March through September began the fifth year. The tame pastures were fertilized annually at 112-25-46 kg/ha of elemental N-P-K. All trees were pruned in two stages to 3.7 m height by plantation age 10 years, at which time the 3.7 x 3.7 m pasture plantations were thinned to about 373 trees/ha (half the original density) because excessive shading was decreasing forage yields. The pastures were burned annually to control insects and diseases in the grasses and to remove pine needles so the grasses would not be smothered.

After 20 years the pines growing in fertilized pastures were larger than those growing in unfertilized native vegetation and had produced about 30 percent more wood (Table 1) even though survival through age 10 had been better in the native range (84 vs. 64 percent) than in the pastures. The trees were tallest in the original 3.7 m spacing but the diameters were greatest at wider spacing. Wood yields were also greatest in the denser planting. Species of pasture grass had little influence on tree response.

Accumulated liveweight gains over 150 years by yearling cattle grazing the tame pastures was 3,900, 2,300, and 1,500 kg/ha from pastures with no trees and with slash pine planted at 6.1 x 6.1 and 3.7 x 3.7 m, respectively. Pensacola bahiagrass was the most shade tolerant as evidenced by its persistence and higher yields. Liveweight cattle gains were 2,100, 2,600, and 3,000 kg/ha for Coastal bermudagrass, dallisgrass, and Pensacola bahiagrass, respectively.

13.3.3. Grass yields under pines

Since the previous study did not measure forage yields, another study

Table 1. Growth responses of slash pine after 20 years when planted at 3.7 x 3.7 and 6.1 x 6.1 m in tame pastures and native vegetation.

	Pasture		Native vegetation	
	3.7 x 3.7m	6.1 x 6.1m	3.7 x 3.7m	6.1 x 6.1m
Height (m)	19[a]	17[b]	16	16
Diameter (cm)	30[a]	33[b]	20[a]	23[b]
Basal area (m^2/ha)	20[a]	13[b]	21[a]	9[b]
Volume (m^3/ha)	181[a]	91[b]	146[a]	64[b]

[1/] Pairs of means marked with different letters are significantly different ($P \leq 0.05$).

was established in 1962 to determine yields of bahiagrass, Coastal bermudagrass and dallisgrass in slash pine planted at 3 x 3 m (17). The pines were 5 years old when grasses were planted and annual fertilization thereafter was at rates of 56, 112, 224, 336, and 448 kg/ha of N with ratios of N-P-K being similar to the previous study. Yields were measured every 6 weeks during the next 4 years.

The year after establishment, grass yields were good (Figure 1), but only about half of open pasture yields. Thereafter yields decreased linearly as canopy closure and shading increased. By plantation age 10 (1966) grass yields were 1,000 kg/ha or less with bahiagrass remaining the most productive. Pine basal area had increased to 24.7 m^2/ha.

Fertilization rates at 224 kg/ha and higher produced the most grass during the first two years. However, as the canopy closed, fertilization rates of 56 to 112 kg/ha were generally better than the higher rates. As in the previous study, bahiagrass was the most shade tolerant and Coastal bermuda was the least. Pines responded equally to all fertilization rates; therefore it appears the lowest rate adequately met the nutritional needs for the more rapid growth of slash pine.

13.4. LATER RESEARCH IN AGROFORESTRY

13.4.1. Combining pines and pasture for hay and grazing

Since the earlier pine-pasture-grazing study had delayed grass planting and grazing for 4 years while the trees reached a cow-resistant size (about 3 m), a 1967 study tested the feasibility of planting pines and pasture grasses simultaneously, producing hay while the trees were small, and grazing after the trees were fully resistant to injury (29). Slash pine was planted in wide-row configurations, 3.0 x 14.6 and 4.9 x

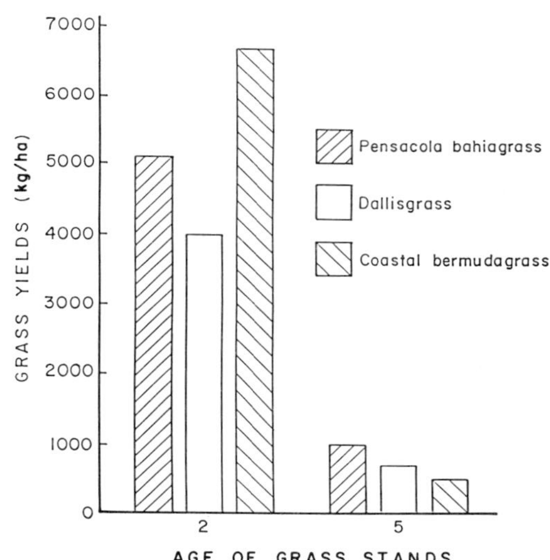

Figure 1. Yields of tame pasture grasses planted under 5-year-old slash pine during the second and fifth years after planting show the influence of canopy closure.

9.1 m, in pastures recently sprigged with Coastal bermudagrass or seeded to Pensacola bahiagrass. Pastures were fertilized and cut four or five times each year for hay during the first 3 years, and, then were grazed for 3 years.

The wide-row spacings easily permitted hay operations and only a few trees were killed while mowing the pastures. Although the tree rows and turning areas removed 5 to 8 percent of the area from production, hay yields the third year of 15,680 kg/ha were about normal for this locale. Liveweight gains by yearling cattle were similar to those from open pastures in the earlier study.

Slash pine planted in these fertilized pastures survived well and grew rapidly as in the previous study. After 6 years, 83 percent of the trees survived and averaged 6.5 m in height and 13.2 cm in diameter at breast height.

13.4.2. Pines planted in established pasture

Some form of mechanical treatment is generally used to prepare sites for planting pines in the southern United States. In the previous studies on combining pines and pastures, the pines had been planted on recently prepared sites. Opportunities for planting pines in a dense pasture sod still needed evaluation. In 1967, slash pine was planted at 3.7 x 3.7 m in dense pasture sod and in clean-cultivated sites with and without fertilization (27). The pasture sod consisted of a mixture of Pensacola bahiagrass, dallisgrass, Coastal bermudagrass, and white clover. The pasture sod was mowed periodically each year, the cultivated treatment was disked each year for weed control, and fertilizer (112-25-46 kg/ha of N-P-K) was applied annually.

The pine seedlings became well established the first year (96 percent survival) and after 5 years 86 percent remained alive. Growth of the planted pines was not affected by site condition at time of planting nor by subsequent cultivation or fertilization. After 5 years the trees averaged 4.8 m in height and 8.9 cm in diameter.

This study, related experience, and other plantings showed that slash pine can be successfully planted even into a thick cover of ground vegetation. However, rainfall preceding and following this planting was good and the seedlings were not apparently climatically stressed. In contrast, researchers in New Zealand have determined that Monterey pine (Pinus radiata D. Don) seedlings are very susceptible to competition from ryegrass (Lolium spp.) pasture or weedy farmland and needs initial and, possibly, later elimination of competition (46, 53).

13.4.3. Seeding bahiagrass during site preparation

Most of the native forage plants in the pine flatwoods ecosystem are low in nutrients and are poorly digestible by cattle (18, 30). Establishing some better forage plants among these native species would improve forage yields and quality. Seeding a tame grass that requires little maintenance, in conjunction with mechanical site preparation for planting pines, could provide improved forage at a reasonable cost. Experience with Pensacola bahiagrass indicated it could be used for this purpose because it can withstand periods of drought and flooding, persist without annual fertilization, tolerate considerable shading, be established by seeding, respond to fertilization, and tolerate close grazing. A study on the Osceola National Forest in north Florida tested

its ability to become established with minimum soil disturbance and no
fertilizer or lime (32). In March 1972, a 76 ha area was sheared
(root-plowed) and strip-disked on 3.7 m centers. Pensacola bahiagrass
seed were broadcast at 16.8 kg/ha on two 4 ha sites shortly following site
preparation. The strips were row-seeded to longleaf pine in December
1973. A light application of fertilizer (34-15-28 kg/ha N-P-K) was
applied to half of each site in May 1975. Cattle had free access yearlong
and grazed this area very heavily throughout the study.

The seed germinated and established a reasonably good stand of
bahiagrass intermixed among the native plants that survived or
re-established following site disturbance. After 4 years, bahiagrass
covered 22 percent of the ground surface. The major native herbaceous
plants were pineland threeawn (Aristida stricta Michx.) and low-panicum
grasses (Dichanthelium spp.) which covered 12 and 6 percent of the ground
surface, respectively.

The single fertilization increased bahiagrass yields by three times
the first year (3,600 vs. 1,100 kg/ha) and two times the second year
(1,680 vs. 880 kg/ha) compared to unfertilized areas. Longleaf pine seed
germinated nicely and these seedlings began height growth earlier and were
50 percent taller at age 9 with grazing than in ungrazed exclosures (4.6
vs. 3.2m). In spite of heavier mortality with grazing (36 percent vs. 21
percent), stocking on the grazed areas was 2,400 trees/ha at age 11.
Overall, this resulted in a much improved forage condition in an excellent
stand of longleaf pine.

13.5 RESEARCH ON COOL-SEASON PLANTS UNDER PINES

The majority of native forages growing in southern forests are
warm-season plants which become dead or dormant during the winter. This
low-quality forage must be supplemented or replaced by a higher-quality
feed, usually hay or concentrates, to maintain animal health and growth.
In an effort to reduce feed and labor costs, various cool-season plants
have been tested for their adaptability to southern climes, soils, and
forests. Research in Louisiana and Mississippi has focused on some
cool-season grasses and legumes.

13.5.1. Grasses

In 1967 a study was established on the Palustris Experiment Forest in
central Louisiana, where annual precipitation averages 1473 mm, to
determine the productivity of some selected cool-season exotic grasses
under 15-year-old mixed slash and loblolly pines (39). Pine density was
about 2,000 stems/ha. The area was prescribe burned in November and four
varieties (Fawn, Goar, Kentucky 31, and Kenwell) of tall fescue (Festuca
arundinacea Schreb.), all cool-season perennials, were seeded in December
at a rate of 28 kg/ha. Four years after seeding (tree age 19 years),
Kentucky 31 and Kenwell tall fescue yielded significantly more forage than
the other varieties; however, residual native grasses outproduced all the
exotics. Native grass production average 178 kg/ha while Kentucky 31 and
Kenwell averaged 105 kg/ha. After 7 years (tree age 22), average annual
yields for Kentucky 31 and Kenwell tall fescue were 450 kg/ha while the
other fescues had essentially disappeared. These cool-season tall fescues
were higher in crude protein, ether extract, ash, phosphorus, vitamin A
and digestibility than the residual native bluestems (Schizachyrium spp.).
However, total forage yields from the land did not significantly increase

with the addition of these cool-season plants. Consequently, the primary
benefit from seeding exotic grasses under dense stands of pine without
fertilization is the availability of green forage during winter which is
higher quality than native grasses.

In Mississippi, the dry matter yields of ryegrass (<u>Lolium perenne</u> L.)
and Kentucky 31 tall fescue grown under artificial shade decreased for
both species as level of shade increased (Table 2) (54). However,
performance of both species was satisfactory under 50 percent shade and
yields of fescue were almost as high at 75 percent shade. Both species
yielded sufficient seed to reestablish themselves the next growing season.
Establishment of a good stand the first year and regulation of late-spring
grazing to permit seed production are critical to continued persistence of
these species.

The effects of cattle grazing on newly planted slash pine in a Gulf
ryegrass pasture were determined in another study (42). During February
1985, pine seedlings were hand planted on 1.2 x 10.1 m spacings or about
815 stems/ha. The control trees were completely protected from grazing
with electric fences 0.3 m on both sides of the seedling rows. To protect
seedlings in the grazed strips from trampling, a single-wire electric
fence was placed about 0.4 m directly over the pine seedlings. Cattle
grazing was programmed for 3 days per week, 3 to 4 hours each grazing day,
with 33 cows, 28 calves, and one bull. Cattle grazed somewhat underneath
the single-wire fence which reduced the ryegrass competition around the
pine seedlings on the grazed plots while the ryegrass on the controls grew
vigorously and rapidly overtopped the pine seedlings. By June 1985,
seedling survival was 56 percent on the controls and 79 percent on the
grazed plots, but only 7 percent of the mortality could be attributed to
cattle injury. These results illustrate the need for competition control
for good survival of the newly regenerated pine seedlings in ryegrass and
that cattle injury to newly planted pines can be kept low with over-head

Table 2. Dry mass (kg/ha) of ryegrass and tall fescue, under three levels
of shade, on three dates in 1980 (54).

| Percent Shade | Yield | | | |
	Feb	Apr	May	Total
	Ryegrass			
0	880	6691	1551	9122
50	694	5040	1297	7031
75	672	2325	1208	4206
	Kentucky 31 Tall Fescue			
0	831	3573	1455	5859
50	806	2705	857	4368
75	757	2567	702	4026

electric fences in intensively grazed pastures. The results are similar
to those regarding susceptibility of Monterey pine seedlings to
competition from ryegrass pasture (45).

13.5.2. <u>Legumes</u>

In the search for cool-season forages as winter feed along with
native forages, legumes such as subclover (<u>Trifolium</u> <u>subterraneum</u> L.) were
evaluated because they appeared adaptable to forest environments in other
countries such as Australia and New Zealand. Advantages also accrue
because they can persist under heavy grazing, have nitrogen fixation
capabilities, are adopted to harsh environments, grow throughout the
winter and spring in the Southeast, and are prolific reseeders (54).
Other researchers (15, 49) confirm that subclover, crimson clover (<u>T.
incarnatum</u> L.) and tall fescue would be excellent choices for use in
forested settings in the South.
The artificial shade studies in Mississippi mentioned earlier evaluated
the performance of several species and varieties of cool-season legumes as
well as grasses (54). For the 13 cool-season legumes tested, the level of
shade had relatively little effect on germination although some exceptions
were noted (Table 3). However, dry matter yields decreased by 40 percent
or more under 50 percent shade and by 75 percent or more under 75 percent
shade. Exceptions were Nangeela subclover and Berseem clover where
production was decreased only 8 and 12 percent, respectively, by 50
percent shade. In contrast, root yields and the percent of root weight
represented by nodules were not affected by shade. Nodules made up 12 to
15 percent of the root weight. In order to have good nodulation and the
resulting nitrogen fixation it is essential to inoculate the stand with
the proper <u>Rhizobium</u> bacteria (21, 22). Shade generally reduced the
height of the upright growing species. Decumbent species such as the
subclovers showed less variation in height among the levels of shade.
There was a trend for percent dry matter to be lower in all species as
level of shade increased. Plant maturity was delayed 5 to 10 days by
shade which would extend the grazing season.
A cooperative study between the Louisiana Agricultural Experiment
Station and the USDA Forest Service established subterranean clover under
mature pines on three locations in Louisiana (21). Pine timber was
thinned to 0, 9, 18, and 28 m^2/ha of basal area. Residual slash was
piled, the plots were prescribed burned, and clover was seeded in 1981.
Subterranean clover yields were highly associated with canopy cover. In
general, yields were about 25-30 percent less under forest compared to
open areas.
Based on these trials, a 4-ha forested area near Glenmora, Louisiana,
was seeded at 17 kg/ha with three varieties of subterranean clover--
Nangeela, Mt. Barker, and Woogenellup--for winter grazing (10). The slash
pine basal area was about 17 m^2/ha and trees were 23 years old. The site
was prescribed burned during late September 1983. A similar test near
Clinton, Louisiana, also included the Meteora variety. Loblolly pine
overstory was 35 to 45 years old with about 20 m^2/ha basal area. Areas
were seeded with 22 kg/ha subclover seed during the first week of November
following mowing and light discing. Forage production was measured in
mid-May. At both study sites, the Nangeela variety generally produced the
most forage. Woogenellup produced slightly more forage in a couple of
treatments. The Nangeela variety also remained green 2-4 weeks longer
than the other varieties. Therefore, Nangeela seed from Oregon appears to

Table 3. Seedling counts, yield, height and percent dry matter of 13 legumes grown under three levels of shade (54).

Species or variety	Level of shade (%)											
	0	50	75	0	50	75	0	50	75	0	50	75
	Seedlings per 30.5 cm of row			Yield kg/ha	% of yield in full sun		Plant height (cm)			Dry matter (%)		
Trifolium subterraneum L.												
Nangeela subclover	46	49	47	4350	92	49	12	13	9	28	23	25
Woogenellup subclover	30	33	30	4700	59	26	14	15	11	29	25	26
Miss. ecotype subclover	34	34	35	4816	55	32	9	12	10	24	21	26
Tallarook subclover	37	40	42	5833	48	26	14	14	10	25	22	25
Dwalganup subclover	23	32	25	2533	65	24	12	11	9	29	24	29
Trifolium vesiculosum Savii.												
Amclo arrowleaf clover	11	9	13	3600	34	27	59	30	23	24	18	21
Yuchi arrowleaf clover	4	8	6	2417	57	49	51	32	30	22	18	18
Meechee arrowleaf clover	2	4	10	2100	39	23	47	29	22	21	16	17
Trifolium resupinatum L.												
Persian clover	15	16	15	1667	58	36	25	17	12	25	18	20
Trifolium alexandrinum L.												
Berseem clover	8	23	12	1350	88	56	39	31	19	24	17	17
Trifolium incarnatum L.												
Chief crimson clover	87	38	35	5016	57	31	43	26	22	34	26	24
Tibbee crimson clover	28	40	31	3517	62	34	44	25	18	47	23	29

the best variety of subterranean clover for use under pines but Woogenellup is acceptable.

During March 1981 loblolly pine seedlings were hand-planted in an open 4-ha subterranean clover pasture in paired (grazed and ungrazed) areas on a 1.8 x 3.1 m spacing. Pine mortality caused by cattle damage was only 6.3 percent, and survival was not significantly different between treatments. After 2 years, pine heights were somewhat less on grazed plots than ungrazed and were not influenced by the subterranean clover. Therefore, it appears that loblolly pine can be successfully established in subterranean clover pasture. However, it will take additional years of growing pines and subterranean clover on the same site to show any beneficial effects to the pines through nitrogen fixation by the clover.

13.6. SPECIAL CONSIDERATION FOR PLANTED PINES

13.6.1. Planting configurations

Historically, pine plantations have been planted in square or nearly square configurations so as to fully utilize all available space for root and crown spread. However, this also provides maximum influence on understory plants through competition for light, water, and nutrients. Consequently, forage species composition and yields are diminished early in a tree rotation (24, 38). To extend the period of forage production farther into the tree rotation (possibly throughout), fewer trees need to be planted and/or planting configurations need to be altered. Research in south Georgia, central Florida and south Florida has shown that planting pines in double-row configurations offers attractive alternatives for maintaining a desirable density of trees while maintaining relatively open conditions to favor understory plants (33).

The major study (33) in this area compared a density of 1,111 trees/ha planted in three single-row and three double-row configurations (Table 4). For clarification, a (1.2 x 2.4) x 12.2 m double-row configuration implies 1.2 m between trees within rows, 2.4 m between rows, and 12.2 m between pairs of rows. Following double-chopping, slash pine seedlings were planted in January-February 1970. At plantation age 13, the trees and vegetation were samples to determine tree growth and forage yields.

Mortality was higher than usual, averaging 32 percent, due to pitch canker (_Fusarium moniliforme_ Sheld. var. _subglutinans_ Wollenw. and Rienk.) disease throughout the stand, wildfire damage, and a road made by hunters driving across the area (Table 4). Tree heights were typical of natural stands of similar age on these sites with no significant differences among planting configurations. Likewise, diameters were similar among all configurations. Both basal area and volume estimates indicated that the (1.2 x 2.4) x 12.2 configuration had the best timber growth while the (0.6 x 2.4) x 26.8 had the poorest. The double-row configurations maintained the most open canopy and produced more forage than the single-row. Therefore, some double-row configurations seem to offer equal timber yields and continued high forage yields as compared to the more typical planting configuration of 2.4 x 3.7 m (Fig. 2).

13.6.2. Livestock injury to planted pines

Pines are usually not browsed as readily as broad-leafed trees but some types of livestock, such as hogs, can be very damaging. Therefore,

Table 4. Average survival, height, diameter at breast-height, basal area,
and forage yields at age 13 of slash pine planted in single-row and
double-row configurations at 1,111 trees/ha, Withlacoochee State Forest,
central Florida.[1]

Spacing configuration (m)	Survival (%)	Height (m)	Diameter (cm)	Basal area (m^2/ha)	Forage yields (kg/ha)
2.4 x 3.7	61	10.5	14.5	11.6	1223
1.2 x 7.3	68	10.6	13.2	11.2	605
0.6 x 14.6	68	11.1	13.0	12.0	1195
(1.8 x 2.4) x 7.3	67	9.8	12.7	9.1*	1577
(1.2 x 2.4) x 12.2	67	11.0	14.0	13.6	1416
(0.6 x 2.4) x 26.8	74	10.2	10.9	7.6*	2882*
Average	68	10.5	13.2	10.8	1483

[1] Means in a column marked with an asterisk are significantly different
from the control treatment (2.4 x 3.7) at the 0.05 level.

Figure 2. Slash pine planted in double-row configurations, such as this
(1.2 x 2.4) x 12.2 m, grew at normal rates while maintaining open
conditions for good yields of forage plants.

the species of tree and type of livestock are very important in
anticipating animal damage to young trees. Cattle occasionally injure
pines by browsing and trampling (40, 55) but severe damage and mortality
primarily occur as a result of having too many animals for the amount of
forage available or because mineral, water, and supplemental feed sources
are placed or occur within young pine plantations. Young plantations
suffer little damage if cattle numbers are kept in balance with the forage
supply (1, 14, 43).

The most obvious way to prevent injury by animals is to exclude all
grazing. Equally successful is deferring grazing during the first growing
season or up to 18 months after planting pines. When grazing is initiated
within 2 years of planting pines, cattle stocking should be kept low or
about one-half the established capacity as determined by forage
availability. After trees are 1.0 to 1.5 m tall, little injury generally
occurs and stocking can be increased.

Learning to accept some injury to pines is important because not all
injury is damaging to survival and growth. Studies in south Florida (19)
and south Georgia have shown that injury must be severe to greatly affect
young slash pine (25, 26). Known levels of injury were hand inflicted on
slash pine at 6, 18, and 30 months after planting. Foliage removal,
growing shoot removal, and stem bending ranged from 0- to 100-percent in
all possible combinations. Stem bending had little effect on survival and
growth. Only the highest levels of combined defoliation and shoot removal
reduced survival, especially on recently planted seedlings. Height growth
was reduced by high levels of combined foliage and shoot removal,
especially at 6 months after planting. Although slash pine is able to
sustain the usual injuries inflicted by grazing animals, repeated injury
can be very harmful to planted pines. Therefore, frequent monitoring of
young stands is necessary to avoid excessive injury.

13.6.3. Insects and diseases

Accelerating the growth of slash pine through cultivation or
fertilization has on occasion been observed to increase the incidence of
insect and disease attacks (13, 20, 28, 29, 31, 51). The most troublesome
have been southern pine coneworm (Dioryctria amatella Hulst.) and southern
fusiform rust (Cronartium quercuum (Berk) Miyabe ex Shirai f. sp.
fusiforme (Burdsall and Snow)). Both of these pests, particularly in
combination, have contributed substantially to increased mortality in some
plantations. However, in other years and other plantations there has been
no increase in attacks by these pests. Managers should always be alert to
detect insect and disease pests, especially when growing plants in a new
environment or under drastically changing conditions. Pests as components
of most ecosystems should receive attention in agroforestry along with
other considerations (4, Altieri et al., this volume).

13.7. OTHER PINE-PASTURE EXPERIENCES

13.7.1. Practical examples

Many landowners in the South have adopted various forms of management
which integrate grazing with forestry, row crops, and pasture production.
Some have been rather innovative in integrating pasture species into their
forest management program (3). There are many examples of pastures that
were planted to pines in the 1950's under a government-sponsored

conservation effort, the Soil Bank Program. Bahiagrass is still prevalent
in many of these slash pine plantations. For years, Mr. Charlie Dixon of
Andalusia, Alabama, planted bahiagrass under mature pines that he managed
for gum naval stores (turpentine), sawtimber, and plywood in a diversified
forestry, farming, and livestock enterprise (6). He only fertilized the
bahiagrass when fertilizer and cattle prices were right to show a profit.
By being so diversified he always had a profitable crop to market and was
able to have a cash crop annually.

Another landowner in central Georgia produces row crops, cattle,
timber, and gum naval stores (44). In this case, beef cattle graze crop
residues, some tame pasture, and 101 ha of multiple-use timberland which
contains 40 ha that were planted to slash pine and bahiagrass in 1953 and
are fertilized annually and grazed at 1.0 ha per animal unit for about 8
months per year. Over 27 years, calves gained weight at about 0.9 kg/day;
selective thinnings have yielded 54 m^3/ha of pulpwood; 1,250 naval stores
trees have been maintained; and the residual sawtimber averages 19 m in
height, 28 cm in diameter, and 19 m^2/ha of basal area. The owner plans to
sell the sawtimber and replant slash pine into the existing bahiagrass
sod.

In south Alabama on the Conecuh National Forest, a 32.4 ha tract
known as Spicer Field, an old bahiagrass pasture, was obtained by the
Forest Service in a land exchange (34). To prepare the land for planting
pines, a watermelon (<u>Citrullus vulgaris</u> Schrader ex Ecklon & Zeyher) crop
was grown during the summer before planting slash pine at 2.4 x 3.7 m. An
excellent stand of bahiagrass re-established from residual seed during the
first year. Starting in the second year, a leasee was allowed to graze
this area for 6 months (April through October) with 30 cows and their
calves. He applied fertilizer annually and mowed the weeds as needed;
cattle stocking was increased to 2.5 cows per ha until forage yields
diminished in the eighth year. The annual fertilization increased tree
growth and allowed a commercial thinning during the ninth year to open the
canopy for increased forage yields. Pulpwood yields were 30 m^3/ha and
calf gains were about 0.9 kg/day each year.

In 1977, a demonstration was established on the Withlacoochee State
Forest in west-central Florida where pines, pasture, and cattle were
concurrently managed. A 32 ha pasture was double-disced to kill the old
grasses and then seeded to Pensacola bahiagrass. Slash pine was planted
during the winter of 1977-78 at 2.4 x 3.7 m. Cattle began grazing in July
1979 and annually graze the pasture from mid-March to October when they
are placed on native forested range and winter-pasture supplements. This
combination supports about 125 animals, produces excellent tame pasture
with annual fertilization, and supplies nutrients for rapid growth of
pines.

13.7.2. <u>Other research</u>

Research on combining pines and tame pastures spread to other areas
of the world during the 1970's and early 1980's. Various species of pines
and various tame pasture grasses are being tested. For example, the
Louisiana Agricultural Experiment Station in Homer is testing Coastal
bermuda-grass under recently planted and mature loblolly pine on Upper
Coastal Plain sites (7).

Research in the southern hemisphere seems to have concentrated on
using the fast-growing Monterey pine in combination with various grasses
and clovers, primarily white clover and subterranean clover. Some

examples are in New Zealand (46, 47), Australia (2, 47), and Chile (36).
In many of these studies, sheep rather than cattle are being used. Since
Monterey pine is intolerant of competition, especially shading, chemicals
are usually applied in spot or strip-spray treatment for grass or weed
control shortly before planting and as needed during the first 2 years
after planting.

Economic considerations associated with agroforestry have generally
been favorable (2, 23, 47). Small landowners and ranchers that already
have adequate fencing will probably find it more profitable than
ownerships where fences must be built to control animals. Another
important consideration is the quality of the wood being produced.
Fast-growing trees can result in low-quality timber because of structural
strength and tendencies to warp (52). Although many properties of wood
are not affected by fast growth (48, 50), the individual tree species
should be tested closely to determine if structural strength and warping
is a problem in construction or lamination. It appears that slash pine
grown in heavily fertilized pastures will be useful primarily for
pulpwood.

13.8. SUMMARY AND CONCLUSIONS

Research and practical experience from many areas of the world
indicates that animal and wood production by combining pines and tame
pasture may be a practical approach to agroforestry (Fig. 3). Careful

Figure 3. With annual pasture fertilization, combined yields from rapidly
growing planted pines and from liveweight gains by beef cattle make
agroforestry an attractive land management technique.

management is required to reduce possible conflicts that can arise, such as overstocking either with animals or trees.

It is important to note that:

1)   Combining pines and tame pasture permits simultaneous production of substantial beef weight gains and potentially rapid growth of planted pines.

2)   Special attention is needed to assure good stands of trees through proper grazing control and prevention of excessive competition from pasture vegetation.

3)   Trees need to be thinned and/or pruned or planted in widely spaced configurations to maintain forage yields.

4)   Forage species need to be relatively shade tolerant.

In the southeast U.S., Pensacola bahiagrass has proven to be more shade tolerant than Coastal bermudagrass, dallisgrass, or carpetgrass. Bahiagrass also withstands flooding and droughts and it will persist without fertilization. A cultivar of subterranean clover, Nangeela, has shown greater shade tolerance under pines than other domestic pasture legumes. It and some cool-season grasses show promise of being useful in providing improved forage conditions under planted pines during the time of year when most pasture and native grasses are dormant.

With careful management, the combining of pines, pastures, and cattle offers opportunities for multiple-product yields that should be financially beneficial. Landowners, especially those with small, non-industrial landholdings, should consider the alternatives offered by agroforestry in their land management programs.

REFERENCES

1.   Adams, S. N. 1975. Sheep and cattle grazing in forests: a review. J Appl Ecol 12:143-152.
2.   Batini, F. E., Anderson, G. W. and R. Moore. 1983. The practice of agroforestry in Australia. In: D. B. Hannaway (ed), Symp Foothills for Food and Forests, pp 233-246. Oregon State Univ, Series No 2, Corvallis, OR. 383 p.
3.   Biles, L. E., Byrd, N. A. and J. Brown. 1984. Agroforestry experiences in the South. In: N. E. Linnartz and M. K. Johnson (eds), Agroforestry in the Southern United States, pp. 152-158. 33rd Ann For Symp, Louisiana State Univ, Baton Rouge, LA. 183 p.
4.   Brunig, F. E. and N. Sander. 1983. Ecosystem structure and functioning: Some interactions of relevance to agroforestry. In: P. A. Huxley (ed), Plant Research and Agroforestry, pp. 221-247. ICRAF, Nairobi, Kenya. 617 p.
5.   Burton, G. W. and A. C. Matthews. 1949. A study of species, seeding methods and fertilizing practices for use on piney woods ranges. Tech Mimeo Pap 1, Georgia Coastal Plain Exp Sta, Tifton, GA. 22p.
6.   Byrd, N. A. 1976. Forest grazing opportunities. For Farmer 35:8-9, 20.
7.   Clason, T. and W. M. Oliver. 1984. Timber-pastures in loblolly pine stands. In: N. E. Linnartz and M. K. Johnson (eds), Agroforestry in

the Southern United States, pp 127-137. 33rd Ann For Symp, Louisiana State Univ, Baton Rouge, LA. 183 p.

8. Combe, J. 1982. Agroforestry techniques in tropical countries: Potential and limitations. Agrofor Syst 1:13-27.

9. Dacy, G. H. 1940. Four centuries of Florida ranching. Britt Printing Co, St Louis, MO. 311 p.

10. Davis, L. G., Johnson, M. K. and H. A. Pearson. 1984. Subclover in pine forests. In: N. E. Linnartz and M. K. Johnson (eds), Agroforestry in the Southern United States, pp. 89-104. 33rd Ann For Symp, Louisiana State Univ, Baton Rouge, LA. 183 pp.

11. Editors. 1982. What is agroforestry? Agrofor Syst 1:7-12.

12. Fowler, S. H. 1969. Beef production in the South. Interstate Printers Publ, Inc, Danville, IL. 858 p.

13. Gilmore, A. R. and K. W. Livingston. 1958. Cultivating and fertilizing a slash pine plantation: effects on volume and fusiform rust. J For 56:481-483.

14. Grelen, H. E., Pearson, H. A. and R. E. Thill. 1985. Establishment and growth of slash pine on grazed cutover range in central Louisiana. S J Appl For 9:232-236.

15. Haines, S. G., Haines, L. W. and G. White. 1978. Leguminous plants increase sycamore growth in northern Alabama. Soil Sci Soc Amer J 42:130-132.

16. Halls, L. K., Burton, G. W. and B. L. Southwell. 1957. Some results of seeding and fertilization to improve southern forest ranges. Res Pap 78, US Dep Agric, For Serv, Southeast For Exp Sta, Asheville, NC. 26 p.

17. Hart R. H., Hughes, R. H., Lewis, C. E. and W. G. Monson. 1970. Effect of nitrogen and shading on yield and quality of grasses grown under young slash pines. Agron J 62:285-287.

18. Hilmon, J. B. and C. E. Lewis. 1962. Effect of burning on south Florida range. Res Pap 146, US Dep Agric, For Serv, Southeast For Exp Sta, Asheville, NC. 12 p.

19. Hughes, R. H. 1976. Response of planted South Florida slash pine to simulated cattle damage. J Range Manage 29:198-201.

20. Hughes R. H and J. E. Jackson. 1962. Fertilization of young slash pine in a cultivated plantation. Res Pap 148, US Dep Agric, For Serv, Southeast For Exp Sta, Asheville, NC. 14 p.

21. Johnson, M. K., Davis, L. G., Ribbeck, F., Render, J. H. and H. A. Pearson. 1986. Management of subterranean clover in pine forested range. J Range Manage 39:454-457.

22. Knight, W. E., Palmertree, H. D. and V. H. Watson. 1976. Growing subterranean clover in Mississippi. Inf Sheet 1268, Mississippi Agr For Exp Sta, Starkville, MS. 2 p.

23. Knowles, R. L. and N. S. Percival. 1983. Combinations of Pinus radiata and pastoral agriculture on New Zealand hill country: Forestry productivity and economics. In: D. B. Hannaway (ed), Symp Foothills for Food and Forests, Oregon State Univ, Series No 2, Corvallis, OR. 383 p.

24. Lewis, C. E. 1974. Grazing considerations in managing young pines. In: Symp Management of Young Pines, pp 160-170. US Dep Agric, For Serv, Southern and Southeastern For Exp Sta, Southern Area State and Private For, New Orleans, LA. 349 pp.

25. Lewis, C. E. 1980a. Simulated cattle injury to planted slash pine: Combinations of defoliation, browsing, and trampling. J Range Manage 33:340-345.

26. Lewis, C. E. 1980b. Simulated cattle injury to planted slash pine: Defoliation. J Range Manage 33:345-348.

27. Lewis, C. E. 1985. Planting slash pine in a dense pasture sod. Agrofor Syst 3:276-274.

28. Lewis, C. E., Burton, G. W., Monson, W. G. and W. C. McCormick. 1983. Integration of pines, pastures, and cattle in south Georgia, USA. Agrofor Syst 1:277-297.

29. Lewis, C. E., Burton, G. W., Monson, W. G. and W. C. McCormick. 1984. Integration of pines and pastures for hay and grazing. Agrofor Syst 2:31-41.

30. Lewis, C. E., Lowrey, R. S., Monson, W. G., and F. E. Knox. 1975. Seasonal trends in nutrients and cattle digestibility of forage on pine-wiregrass range. J Ani Sci 41:208-212.

31. Lewis, C. E., McCormick, W. C. and W. E. White. 1972. Cultivation, grazing, insects, and disease affect yield of slash pine planted in a sod. Res Note SE-174, US Dep Agric, For Serv, Southeast For Exp Sta, Asheville, NC. 8 p.

32. Lewis, C. E., Monson, W. G. and R. J. Bonyata. 1985a. Pensacola bahiagrass can be used to improve the forage resource when regenerating southern pines. S J Appl For 9:254-259.

33. Lewis, C. E., Tanner, G. W. and W. S. Terry. 1985b. Double vs. single-row pine plantations for wood and forage production. S J Appl For 9:55-61.

34. McKathen, G. 1980. The Spicer Field story. In: D. Child and E. Byington (eds), Southern Forest Range and Pasture Symposium, pp 208-211. Winrock International, New Orleans, LA. 268 p.

35. Nair, P. K. R. 1985. Classification of agroforestry systems. Agrofor Syst 3:97-128.

36. Panaloza, R., Herve, M. and L. Sobarzo. 1985. Applied research on multiple land use through silvopastoral systems in southern Chile. Agrofor Syst 3:59-77.

37. Payne, W. J. A. 1985. A review of the possibilities for integrating cattle and tree crop production systems in the tropics. For Ecol Manage 12:1-36.

38. Pearson, H. A. 1974. Range and wildlife opportunities. In: Symp Management of Young Pines, pp 19-27. US Dep Agric, For Serv, Southern and Southeastern For Exp Sta, Southern Area State and Private For, New Orleans, LA. 349 p.

39. Pearson, H. A. 1975. Exotic grass yields under southern pines. Res Note SO-201, US Dep Agric, For Serv, South For Exp Sta, New Orleans, LA. 3 p.

40. Pearson, H. A. 1976. Botanical composition of cattle diets on a southern pine-bluestem range. Res Note SO-216, US Dep Agric, For Serv, South For Exp Sta, New Orleans, LA. 3 p.

41. Pearson, H. A. and J. P Barnett. 1984. Agroforestry--subterranean clover and pines. Establ Prog Rep FS-SO-1701-3.24, US Dep Agric, For Serv, South For Exp Sta, New Orleans, LA. 43 p.

42. Pearson, H. A. and D. A. Rollins. 1987. Ryegrass pasture for supplementing southern pine native range. Rangelands 9:19-20.

43. Pearson, H. A., Whitaker, L. B. and V. L. Duvall. 1971. Slash pine regeneration under regulated grazing. J For 69:744-746.

44. Peebles, L. O. 1980. Integrated forest and range land use. In: D. Child and E. Byington (eds), Southern Forest Range and Pasture Symposium, pp 212-214. Winrock International, New Orleans, LA. 268 p.

45. Percival, N. S. and R. L. Knowles. 1983a. Agroforestry: Expanding horizons. In: Proc Raukura Farmer's Conf, pp 37-40. Raukura, New Zealand.

46. Percival, N. S. and R. L. Knowles. 1983b. Combinations of Pinus radiata and pastoral agriculture on New Zealand hill country: Agricultural productivity. In: D. B. Hannaway (ed), Symp Foothills for Food and Forests, pp 185-202. Oregon State Univ, Series No 2, Corvallis, OR. 383 p.

47. Reid, R. and G. Wilson. 1986. Agroforestry in Australia and New Zealand. Goddard and Dobson Publ, Victoria, Australia. 255 p.

48. Saucier, J. R. 1981. Effects of fast growth rate on wood quality and product yields. In: Tech Assoc of the Pulp and Paper Industry Ann Meeting Proc, pp 435-457. New Orleans, LA.

49. Sawyer, G. E. J., Jr. 1978. Evaluation of four clovers on a forest soil. In: S. G. Haines (ed), Nitrogen fixation in southern forestry, pp 108-111. Intern Paper Co, Bainbridge, GA. 169 pp.

50. Schmidtling, R. C. 1973. Intensive culture increases growth without affecting wood quality of young southern pines. Can J For Res 3:565-573.

51. Schultz, R. P. 1969. What is being learned from intensive management of slash pine. For Farmer 28:8-9, 16.

52. Senft, J. F., Bendsten, A. and W. L. Galligan. 1985. Weak wood: Fast-grown trees make problem lumber. J For 83:476-484.

53. Tustin, J. R., Knowles, R. L. and B. K. Klomp. 1979. Forest farming: A multiple land-use production system in New Zealand. For Ecol Manage 2:169-189.

54. Watson, V. H., Pearson, H. A., Knight, W. E. and C. Hagedorn. 1984. Cool season forages for use in pine forests. In: N. E. Linnartz and M. K. Johnson (eds), Agroforestry in the Southern United States, pp 79-88. 33rd Ann For Symp, Louisiana State Univ, Baton Rouge, LA. 183 p.

55. Williston, H. L. 1974. Control of animal damage to young pine plantations in the South. J For 72:78-81.

	cattle	2, 4, 5, 53, 118, 121, 128, 162, 195-197, 199, 201, 206
	cauliflower	122
Cavia spp.	guinea pig	55
Cedrela odorata		128, 144, 183, 188
Cedrela spp.	white cedar	2, 191
Cedrela toona		125
Cedrus deodora		128
Ceiba petandra		114, 122
Celtis australis		121
Cenchrus ciliaris		127
Cenchrus spp.		127
Chamaecytisus spp.		32
	chickpea	114
	chiku	117
	chilli	111
	Chinese potato	116
Choloepus spp.	sloth	52
Chromolaena odorata	eupatorium	149
Chrysophyllum albidum		141
Chrysopogon fluvus		127-129
Citrullus vulgaris	watermelon	207
	citrus	117, 141
Coccinella nugtoria	coccinellid beetle	95
Coccinellidae		97, 103, 104
Cocos nucifera	coconut	3, 4, 109, 115, 116, 123, 158
	codling moth	100, 101, 103, 106
Coffea arabica	arabica	124
Coffea canephora	robusta	124
	coffee	2, 66, 123, 124
Cola acuminata	cola	141
Cola nitida	cola	149
Coleoptera		95, 104
Collops malachiid		97
Collops spp.		95
Colocasia spp.	taro	12, 116
Colossoma macropomu	tambaqui	158
Copernicia spp.		156
Coriariaceae		31
	corn (see maize)	
Cornus spp.		128
	cotton	111, 114
	cowpea	28, 79, 116, 146, 148, 190
	cows (see cattle)	129
Crataegus mexicana	tejocote	91
Cronartium quercuum	fusiform rust	206
Cryptomeria japonica		114
Cyamopsis tetrogonoloba		118
Cycadaceae		31
Cynodon dactylon	bermuda grass	196-199, 208

	leafhopper	100-103
Leopoldian spp.		157
Lespedeza striata	lespedeza	196
Leucaena leucocephela		3, 28, 78, 122
Leucaena spp.		2, 4, 14, 16, 28, 32, 33, 34
	litchi	117
	loblolly pine	200, 204, 207
Lolium perenne	ryegrass	199, 201
Lonchocarpus spp.	barbasco	182
	loquat	117
Lovoa trichilioides		143
Lupinus spp.	lupine	99
	lycosid spider	96, 98, 99
Macaranga barteri		149
Macrodactylus spp.	scarab	95, 97, 99
Madhuca latifolia	mahua	126
	maguey	94
	maize	28, 33, 74, 76, 91, 92, 94, 97, 99, 106, 111, 114, 116, 121, 146, 147, 148
Malachiidae		95
Mangifera indica	mango	117, 121, 126, 188
	mangrove	3, 118, 129
Manicaria spp.		157
Manihot esculenta	cassava, manioc	12, 68, 74, 76, 78, 116, 141, 143, 146-148, 181, 183, 185, 186, 190
Mauritia flexuosa	buriti or morche palm	156, 157
Maximiliana spp.		157, 161
Maytenus emarginatum		120
Mazama americana	brockett deer	48, 53
Mesopis emini		125
	mesta	114
Metroxylon sagu	sago palm	12, 156
Michelia doltsopa		114
	millet	74, 76, 111, 114
Mimosa spp.		33, 37, 119
	monkey	48
Moringa oleifera		121
Morus serrata		121
Musa spp.	banana	12, 13, 111, 116, 117
Musanga cecropioides		149
	mushroom	76
Myricaceae		31
Myrmecophaga tridactyla	giant anteater	48
	nangeela	208
Nauclea diderrichii		143, 144

	nectarine 117
Nypa fructicans	nipa palm 156
Oerocarpus spp.	157
	olive 3
Orbignya cohune	cohune palm 161
Orbignya phalerata	babassu 155, 159-164
Orbignya spp.	156, 157
	orchid 118
Orius spp.	95, 97, 99
Ougeinia spp.	34
	paddy rice 111, 113, 114, 116, 121
	palm 6, 51, 68, 117, 118, 158, 163, 165
	palm weevil 48
Pandanus spp.	2
	pangolin 48
Panicum maximum	Guinae grass 116
Parkia spp.	4, 33, 34
Paspalum dilatatum	dallisgrass 197, 199
Paspalum notatum	Pensacola bahiagrass 196-199, 207, 208
Pennisetum typhoides	118
Pentaclethra macrophylla	141
Periserianthes spp.	34
	phalsa 118
Phaseolus aconitifolius	118
Phaseolus radiatus	mung 118, 122
Phaseolus spp.	frijol 92
Phoenix sylvastus	126
Phytelephas macrocarpa	yarina 189
Picea smithiana	128
	pigeon pea 114
	pig, hog 53, 67, 129, 158, 204
	pine 2, 195, 198
	pineapple 13, 114, 116, 117, 184-186
Pinus caribaea	146
Pinus elliottii	slash pine 196, 197, 199-207
Pinus oocarpa	146
Pinus palustris	longleaf pine 196, 200
Pinus radiata	Monterey pine 199, 202, 207
Pinus roxburghii	128
Pithecellobium spp.	34
	plantain 190, 191
	plum 117, 121
	pomegranate 117
	2
Pometia spp.	33
Pongamia spp.	
Populus spp.	poplar 113, 114
Poraqueiba sericea	umari 184-186, 189
	porcupine 90
	potato 113, 114, 121, 122
	poultry 129, 158

Pourouma cecropiaefolia	urilla	185
Pouteria caimito	caimito	189, 191
	predaceous beetle	99
Prosopis alba		34
Prosopis cineraria		25, 32, 118-120, 126, 127
Prosopis julifora		34, 120, 122
Prosopis pallida		34
Prosopis spp.	algaroba, mesquite	33
Prunus persica	peach	91, 117, 121
Prunus spp.	capulin	91, 92, 94
Pseudotsuga menziesii	Douglas-fir	50
Pterocarpus dalbergoides		114
Pterocarpus santallinum		123
Pterocarpus spp.		34
	pumpkin	74
Pyrus malus	apple	91, 100-104, 117, 121, 126
Quararibea cordata	zapote	191
Quercus incana		128
Quercus spp.	oak	114
	rape seed	114
	raphia	157
Raphididae		103
Rhamnaceae		31
Rhizobia spp.		11, 202
Rhizophora conjugata	mangrove	129
Rhizophora mucronata		129
	rice	190
Rigidoporous lignosus	fungus	145
Robinia spp.	locust	33
Rosaceae		31
Sabal spp.	sabal palm	161
Saccharum spp.	sugarcane	12, 13, 114
	saffron	118
Salvadora oleoides		120
Samanea spp.		33, 34
Santalum album		114
Scheelea spp.		157, 161
Schima wallichii		114
Schizachyrium spp.	bluestems	200
Sehemia spp.		126
	senji	118
	sesamum	111, 118
Sesbania grandiflora		3
Sesbania spp.		23, 33
	sheep	4, 118, 128, 129
Shorea robusta		114, 128
	silk moth	118
Socratea spp.	shebon	189
Sorghum vulgare	jowar	74, 114, 116, 118, 122
	soybean	76, 114, 116
Spondias dulcis	taperiba	191